Keeping the Heat In

Cataloguing in Publication Data

Main entry under title :

Consumer's guide : keeping the heat in

New ed., 1990, reprinted 1996, 1997
Issued also in French under title: Guide du consommateur, emprisonnons la chaleur.
Cover title.
ISBN 0-662-19781-X
Cat. no. M92-1/1992E

1. Dwellings -- Energy conservation.
2. Dwellings -- Insulation.
3. Insulation (Heat)
I. Canada. Natural Resources Canada.
II. Title: Keeping the heat in.

TJ163.5C66 1995 693.8'32 C94-702167-2 rev.

To obtain additional copies of this publication, refer to page 107.

Aussi disponible en français sous le titre : «Emprisonnons la chaleur»

Table of Contents

Introduction

Vhat Is Retrofitting?

etrofitting a house is simply upgrading it so that it ill "keep the heat in". This means adding insula-on, caulking, weatherstripping, improving or re-lacing windows and doors, and improving the heat-ng system. Retrofitting also means including en-rgy-efficiency measures in all your renovation and epair activities. In our climate, retrofitting usually nakes a lot of sense.

'his Book

. tells you how to go about retrofitting your home. deals with houses of all kinds, in all parts of Can-da. It does not deal with apartment or commercial uildings, although some of the information may be seful.

'his book is designed to serve both the experienced o-it-yourselfer and the novice willing to give it a y. It is also a useful consumer guide for homeown-rs who intend to hire a contractor to do retrofit vork.

f you rent, this book features many low cost meas-res that will save you money and make the house nore comfortable. Be sure to read Chapters 3,7, and , which look at air sealing, windows and doors, and ow to operate your house. You may want to make p a list of recommended activities, and approach ne landlord with your suggestions. If the landlord ays the heating bill, the benefits are obvious. If you ay the heating bill, the landlord gets a happy tenant nd a more valuable house that is protected from noisture damage.

Vhy Retrofit?

inergy Efficiency

'erhaps most importantly, retrofitting a home costs ess than producing new energy supplies to heat it. 'ully fifteen per cent of Canada's annual energy use oes to heat our homes, and this comes mostly from on-renewable resources such as oil and gas.

Comfort

A well-insulated, air-sealed house is a comfortable home! An insulated, tighter house is also a much quieter house, and there is less dust and pollen to worry about.

A Sound House

By considering energy efficiency as part of your maintenance and repair activities, your house will be in better shape. Because you have considered air and moisture control, your work should last longer and look better.

Save Money

Energy efficiency is one of the best investments you can make today, paying tax-free dividends immedi-ately in the form of lower heating costs. Home insu-lation is better than just about any other low-risk, long-term investment you can make.

All this aside, the investment is still a good one — since **it is an investment in the conservation of our valuable energy supplies** and this means we all get the environmental benefits of conservation.

Retrofit Opportunities

What retrofit strategies are best for you? You will have to determine what shape your house is in and what can be done to improve it.

Check the interior and exterior for signs of moisture damage and structural problems, maintenance and repair needs, renovation opportunities, the level and condition of insulation, and air leakage paths. Some utilities offer an audit service to help you assess your needs; call your local utility to see if this serv-ice is available. Some contractors also offer assess-ments; check your *Yellow Pages*.

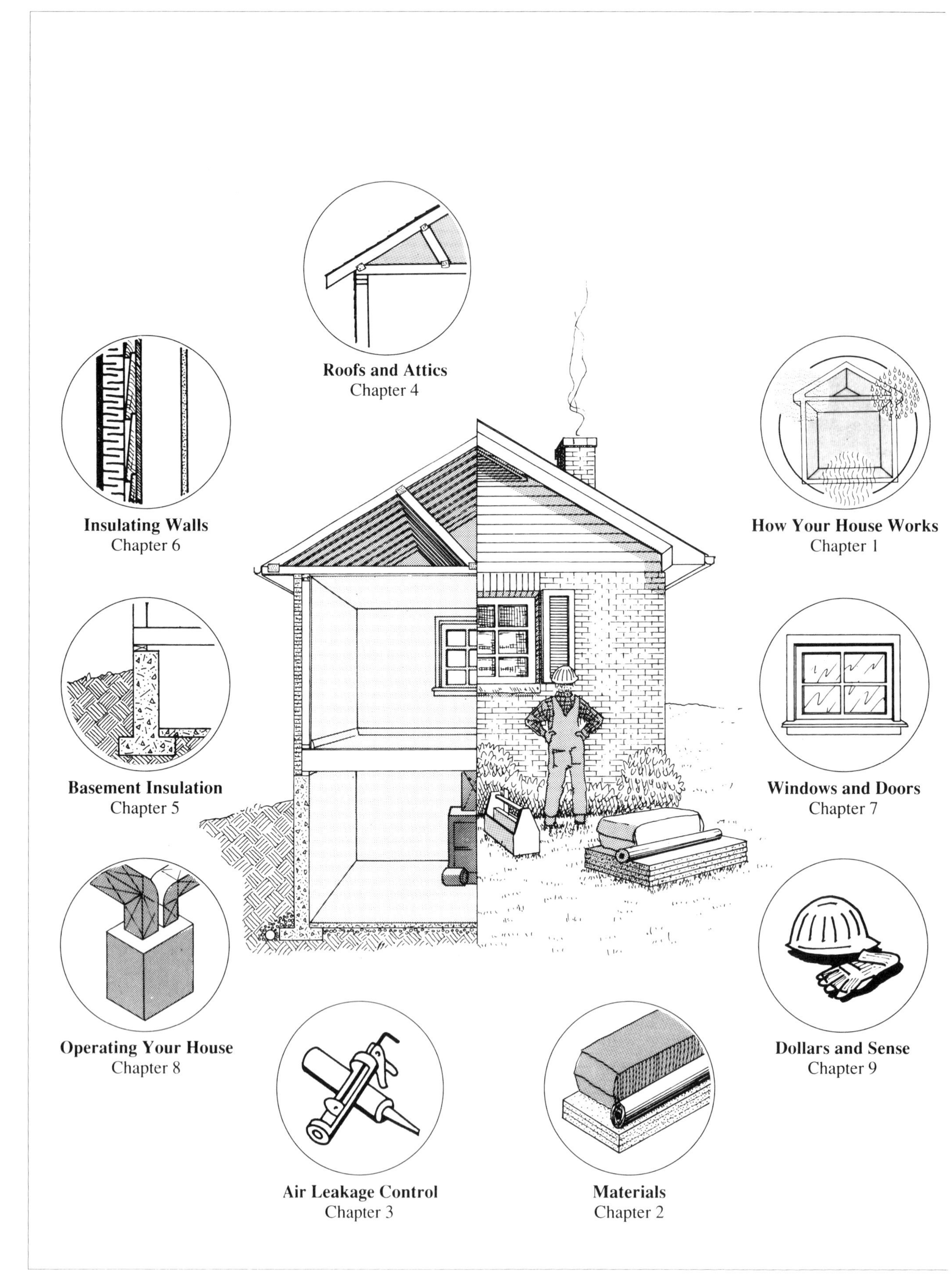
Roofs and Attics
Chapter 4
Insulating Walls
Chapter 6
How Your House Works
Chapter 1
Basement Insulation
Chapter 5
Windows and Doors
Chapter 7
Operating Your House
Chapter 8
Dollars and Sense
Chapter 9
Air Leakage Control
Chapter 3
Materials
Chapter 2

ach house is unique, but some general statements an still be made about retrofit opportunities.

Virtually all houses will benefit from **air leakage control**. Weatherstripping and sealants will stop drafts, save money, improve comfort, and protect the structure. See Chapters 3 and 7 for details. You should also consider moisture control and ventilation to reduce the chance of condensation problems.

Many houses will benefit from a complete **heating system tune-up**. This should analyze and correct any problems with the furnace or boiler, the distribution system, and the controls. Upgrading or replacing the unit with a high-efficiency model will provide substantial savings. See Chapter 8 for a summary.

Insulate a poorly insulated attic. If there is less than 150 mm (6 in.) of insulation in the attic, it will be worthwhile putting more in. It is important to provide a good air seal first. See Chapter 4 for details.

Insulate an empty frame wall. If there is no insulation in a frame wall, it is worthwhile blowing in insulation to fill the cavity. See Chapter 6 for details.

Insulate the basement. Basements are areas of significant heat loss in most houses. If the insulating can be combined with damp-proofing on the exterior or finishing the inside, it will be doubly worthwhile. See Chapter 5 for details.

Make the most of repair and renovation work. Almost all repairs and renovations you do around the house can have an energy efficient component piggybacked onto the work. You will find useful ideas throughout the book.

Other opportunities for retrofit may be right for you. If you think that any part of your house can be made more energy efficient, go to the specific chapter for details. Chapter 8 has a lot of good suggestions for other ways to save energy around the home.

Iow to Use This Book

Everyone should read Chapters 1, 2, and 8. This will rovide important background information on retrofit techniques and materials. Read other chapters as required for specific details. Why not skim through each chapter to see what might be right for your house and renovation plans? Remember, improving the energy efficiency of your house is an ongoing process. Mostly it is accomplished bit by bit as you work on your house over the years. Keep this book around as a handy reference.

Some Important Things to Keep in Mind

House as a System

We have gained a lot of experience with retrofit work over the last decade and a half. One of the most important lessons we have learned is that a house works as a system. Each part of the house is related to all other parts, and making a change in one place causes an effect elsewhere.

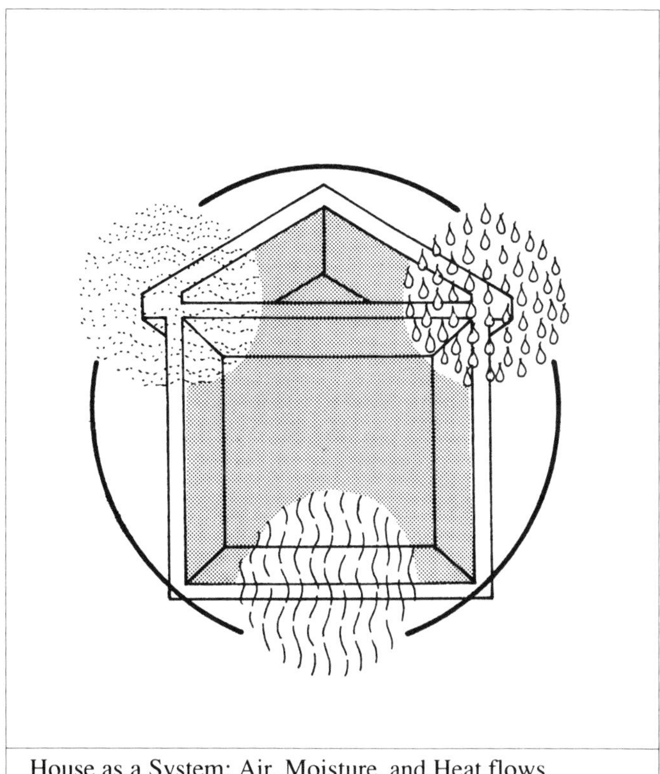

House as a System: Air, Moisture, and Heat flows all interact

There are many forces at work in a house; structural loading, the effects of wind and weather, flows of moisture, heat, and air. These must be kept in the right balance. Adding insulation, air barriers and vapor barriers can effect moisture conditions, ventilation, and combustion air. Chapter 1 discusses this in detail and should be read by everyone.

Health and Safety

Approach any work around the house with your health and safety in mind. Most health and safety rules amount to using common sense around ladders and tools, and when working in cramped and stuffy conditions such as attics. Also, insulation and sealing materials must be handled and installed with care. Chapter 2 describes some general health and safety considerations. Chapters 4 to 7 describe measures specific to working in different areas of the house.

Codes and Standards

Materials specifications, installation procedures, and construction techniques are normally spelled out in codes and standards. Typically, these concentrate on health and safety issues, such as ventilation and fire safety requirements. Each province, and most municipalities, have jurisdiction over the building code. The information in this book, written for all Canada, is general in nature. Local codes should be followed. Check with your local municipal office and building inspector.

Do It Yourself or Hire a Contractor

Much of the work can be done by the do-it-yourselfer with a few special tools and the right materials. The cost savings and job satisfaction can both be high. If you take the trouble to do the job properly, the results should be excellent. Make sure you consult this book carefully. It is designed to meet the needs of both the experienced and inexperienced "do-it-yourselfer".

Some types of retrofit are best done by a contractor with specialized equipment and experience,or you may just prefer to have someone do the work for you. You are far more likely to have excellent work done if you choose a contractor carefully and take an active interest in the work. The more you know, the better. This is especially important if you are hiring a contractor to do general renovations and you want to include energy-efficiency as part of the work.

1 How Your House Works

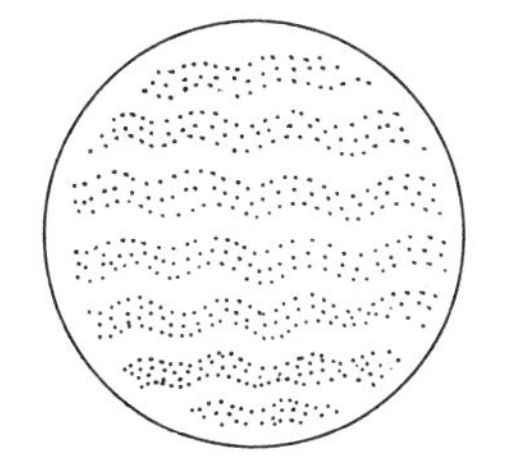

Part III
Control of
Air Flow

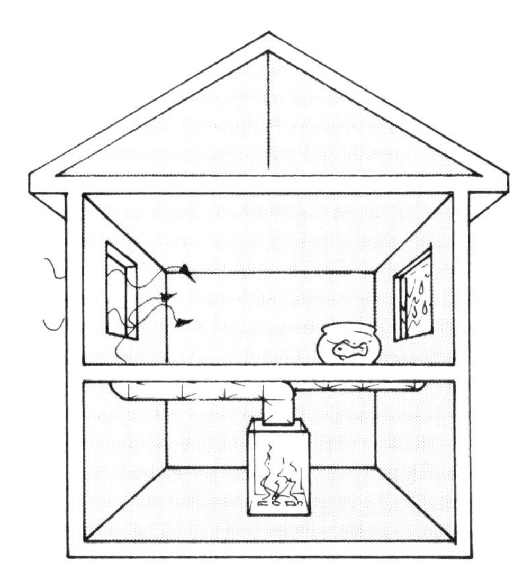

Part IV
Control of
Moisture Flow

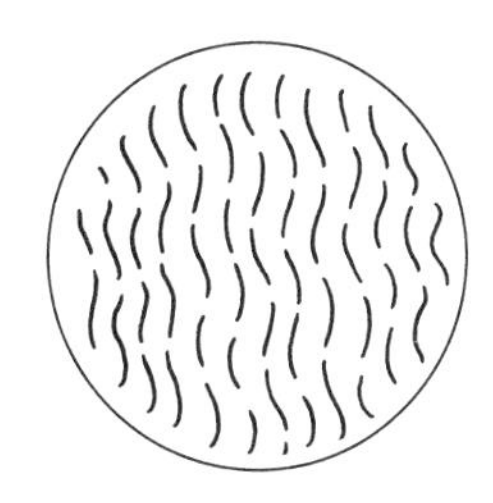

Part II
Control of
Heat Flow

Introduction

It is important to understand how your house works before starting any retrofit work. This will ensure that the job will meet your expectations and that you won't be causing new problems while solving old ones.

This chapter will describe the basic building science that is important for successful retrofit work. It will explain how building science principles are used successfully to control the flow of heat, air and moisture and why these should be considered together.

Part I The Basics

House Performance

We expect our homes to provide shelter from the sun, rain, wind and snow, and we expect them to keep us warm and comfortable. We also expect that they will be sturdy and durable.

A number of factors work together to meet these needs. These include the building shell, the outside environment, the mechanical system, and ourselves, the occupants. This book is mainly about improving the performance of the house envelope.

The Building Envelope

The building envelope is the shell of the house that protects us from the elements; it includes the basement walls and floor, the above-grade walls, the roof, and the windows and doors.

We expect a lot of the envelope; it must provide structural support for the walls and roof, protect the structure from deterioration, and allow for natural lighting of the interior and a means of getting in and out. Finally, the envelope must separate our warm and comfortable controlled indoor environment from the weather outside.

To maintain our indoor environment, the envelope must control the flow of heat, air and moisture between the inside of the home and the outdoors.

The Envelope and Heat Flow

As part of the controlled indoor environment, we add a heating system in order to overcome the cold Canadian winters. We try to build our homes so that we don't heat the Great Canadian Outdoors. We try to keep the heat in!

But heat will move wherever there is a difference in temperature. **Basically, heat flows from areas of warmth to areas of cold.**

Many people believe that because hot air rises, most heat loss will be through the ceiling. Not necessarily so. **Heat moves in any direction** — up, down or sideways — as long as it is moving from a warm spot to a colder one. A heated room over an unheated garage will lose heat through the floor. Similarly, heat loss can occur through walls — in the basement or crawl space, as well as above the ground. Heat moves to the cold. It's the envelope's job to control the flow of heat between the indoors and outdoors.

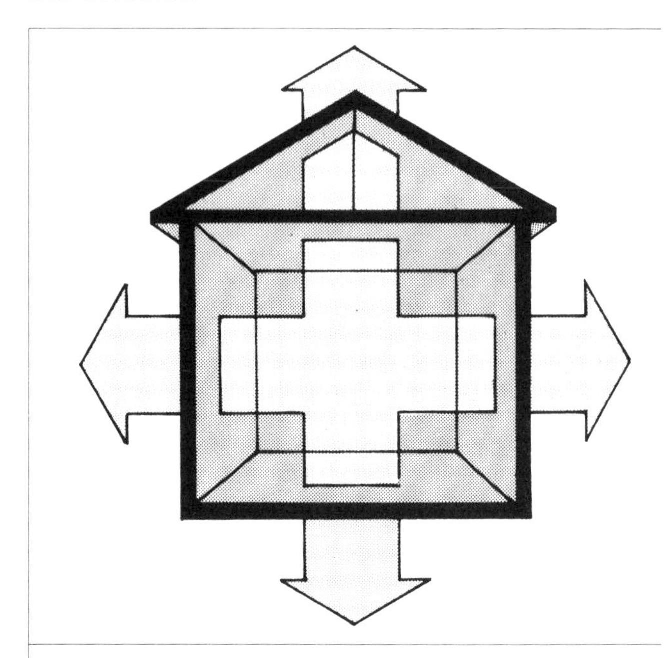

Heat moves out of the house in all directions.

How Does the Heat Flow?

Heat flows in three distinct ways. In a part of the house envelope, such as a wall, heat can be moving in one, two or all three ways at the same time.

- **Conduction**. Heat can be transferred **directly** from one part of an object to another part by the particles bumping into each other. For example, the heat from a cast iron frying pan is transferred to the handle and eventually to your hand. Some materials conduct heat better than others, depending on the structure of the material. Insulation works by reducing heat flow with tiny pockets of air, which are relatively poor conductors of heat.

- **Convection**. Heat can also be transferred by the **movement** of a fluid such as water or air. In an uninsulated wall space, for instance, air picks up heat from the warm wall, and then circulates to the cold wall where it loses the heat. Some heat also transferred by the mixing of warm and cold air.

Radiation. Any object will **radiate** heat in the same way that the sun radiates heat. When you stand in front of a cold window, you radiate heat to the window and so you feel cold, even though the room temperature may be high.

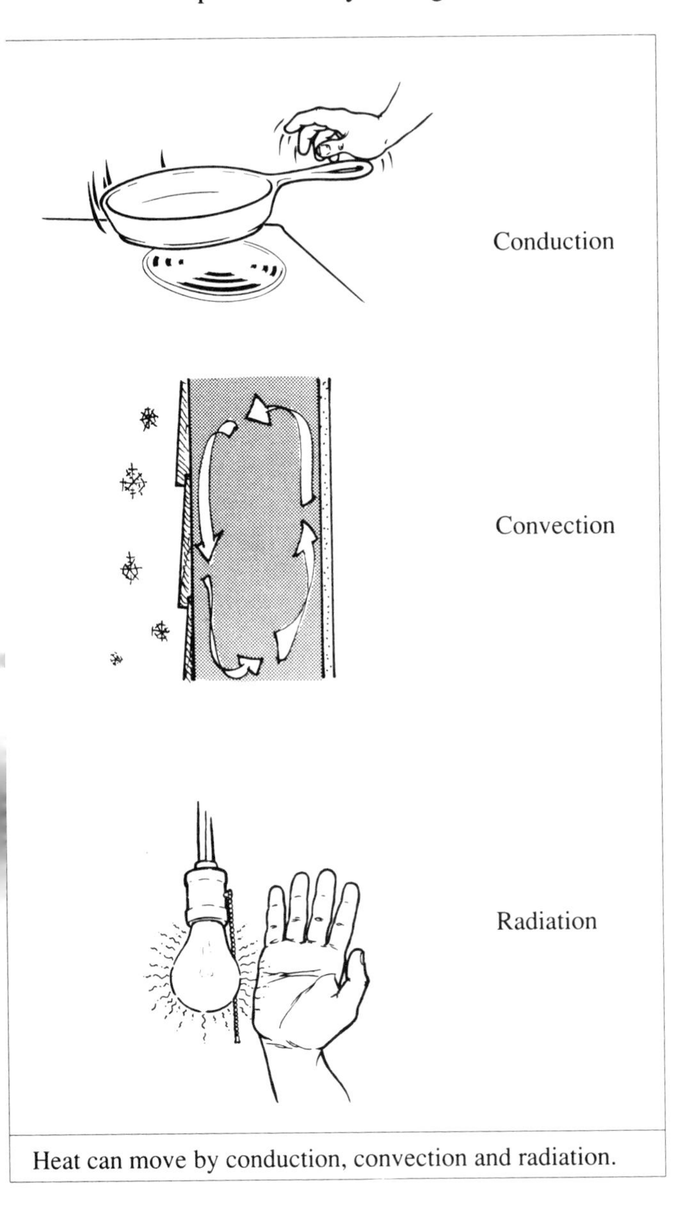

Heat can move by conduction, convection and radiation.

The Envelope and Air Flow

Uncontrolled air flow through the envelope can be a major source of heat loss and can lead to other problems. Since warm air can carry large amounts of water vapor, air flow is also the main means by which moisture is carried into the envelope.

Under winter conditions, air is forced through the building envelope. Air moving out carries heat and moisture, while air moving in brings uncomfortable drafts and dry winter air.

For air to move from one side to the other, there must be a hole in the envelope and a difference in air pressure between the inside and outside. The difference in air pressure can be caused by any combination of:

- wind;
- a temperature difference creating a 'stack' effect in the home; or
- combustion appliances or exhaust fans.

- **Wind Effect.** When wind blows against the house, it creates a high pressure area on the windward side and air is forced into the house. There is a low pressure area on the leeward (and sometimes other) sides where air is forced out.

- **Stack Effect**. In a heated home, less dense warm air rises and expands, creating a higher pressure area near the top of the house. Air escapes through holes in the ceiling and cracks around upper storey windows. The force of the rising air creates lower pressure near the bottom of the house and outside air rushes in through cracks and openings around the lower floors.

- **Combustion and Ventilation Effect**. Appliances that burn fuels such as wood, oil or natural gas, need air to support combustion and provide the draft in the chimney. Open chimneys and fireplaces tend to exhaust lots of air. This air must be replaced, and outside air will be drawn in through the envelope. This is why people often notice that the room becomes drafty when there's a fire in the fireplace.

Ventilation fans in the kitchen and bathroom, central vacuum systems, stove-top grills, clothes dryers, and other exhaust fans also cause this effect.

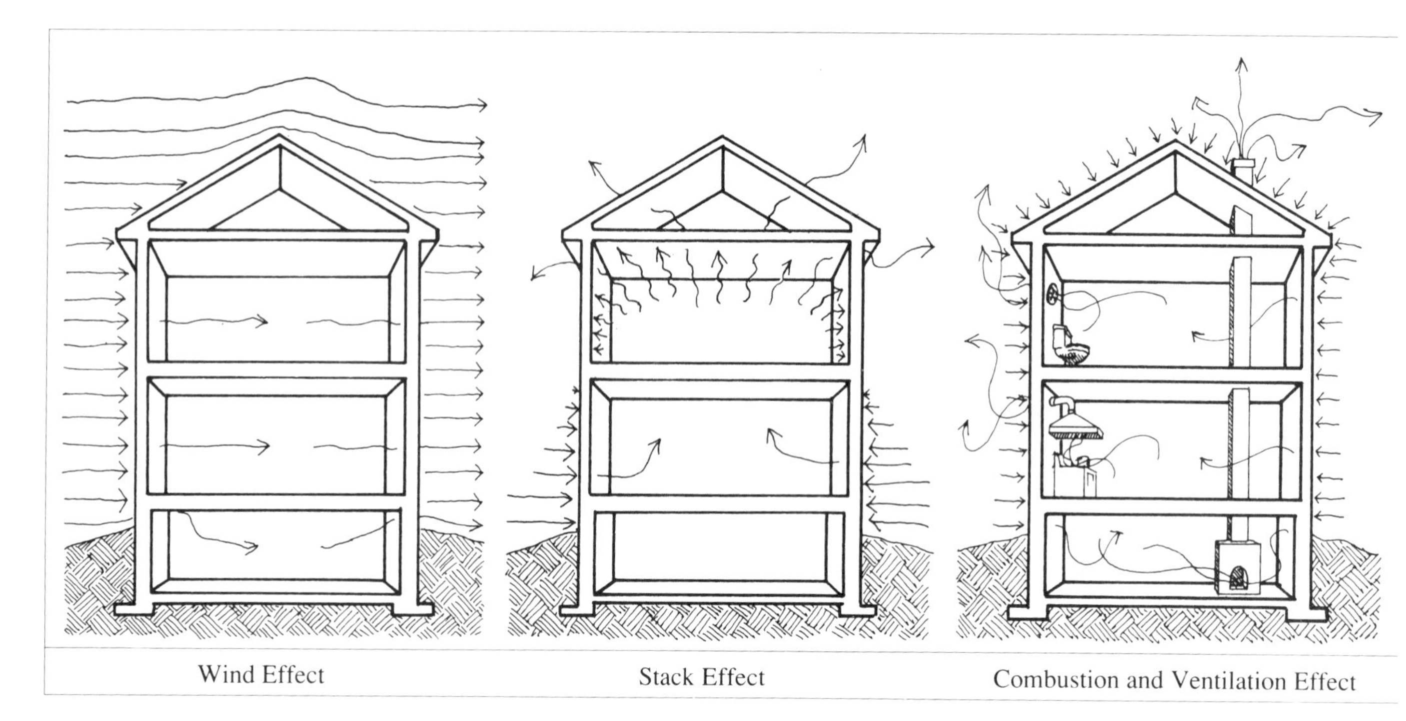

The Envelope and Moisture

Moisture can cause concrete to crumble, rot wood and cause paint to peel; it can damage plaster and ruin carpets. In its many forms, moisture is a major cause of damage to building components.

Moisture can appear in the form of a solid, a liquid or as a gas (water vapor). Moisture can originate from outside, as groundwater in the soil, as ice, snow, rain, fog and surface runoff, OR from inside in the form of water vapor produced by the house occupants and their activities, such as washing, cleaning and cooking, or the use of humidifiers.

In its different forms, moisture can move through the envelope in a number of ways.

- **Gravity**. Water running down a roof or condensation running down a window pane shows how gravity causes water to move downward.
- **Capillary Action**. But water can also move sideways or upwards by capillary action. Capillary action depends on the presence of very narrow spaces, as with lapped siding or porous materials such as concrete or soil (think of how a paper towel absorbs water).
- **Diffusion.** Water vapor can also move directly through materials by diffusion. Diffusion depends on a difference in water vapor pressure and the material's resistance to this pressure.
- **Air Movement.** As water vapor, moisture is carried by moving air, for example, where there is air leakage through a crack in the house envelope.

Far more moisture can be carried by air flow through a small hole in the envelope than by diffusion through the building materials.

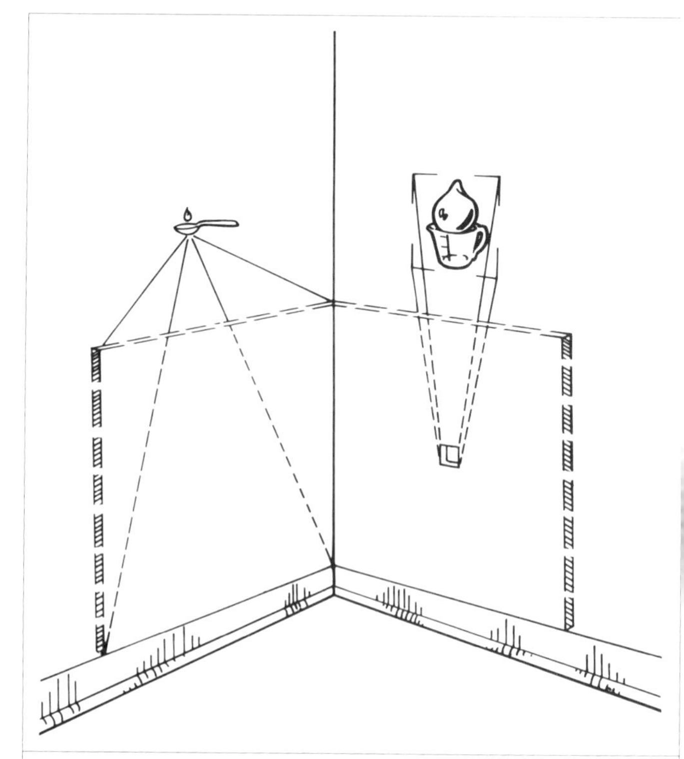

One small air leak can let through 100 times as much moisture as will travel by diffusion over a much larger area.

Condensation

Water vapor becomes a problem when it condenses into liquid water. This happens at the point of 100% relative humidity when the air can't hold any more water vapor.

Condensation occurs when moist air meets a cool surface.

A typical example is condensation on windows. When the air contacts the cold window it loses heat. Now that the air is cooler, it can no longer hold all the water vapor and some condenses out onto the surface of the window. If the window is extremely cold the condensation will appear as frost. A single-glazed window will be colder than a double-glazed window and so condensation will appear under conditions of lower humidity in the room. Condensation is more likely to occur in humid areas of the house such as the kitchen and bathroom.

The House as a System

While this book concentrates on improvements to the house envelope, it's important to remember that the house operates as a system. All the elements of a house, the environment, envelope, mechanical systems, and occupant activities affect each other and the result affects the performance of the house as a whole. Understanding these relationships is the secret to avoiding problems.

For example, reducing air leakage makes the house more comfortable and protects the envelope from moisture damage. But it will also increase humidity levels since less water vapour escapes. This can mean an increase of condensation on windows.

The lesson here is that a change to one component of the house can have an immediate effect on another component. The combined effect of many small changes over time can also affect the balance of the system.

Before beginning any retrofit work, it's a good idea to review what's involved and to understand what other aspects of the house may be affected. Thinking things through and careful planning in the early stages of work will prevent unpleasant surprises and ensure that the work meets your expectations.

Further information on the effect that retrofit work has on the house as a system is included in the balance of this chapter and in Chapter Three and Chapter Eight. If you have any questions, talk to an expert or a contractor who is familiar with the principles of how a house works as a system.

Part II Control of Heat Flow

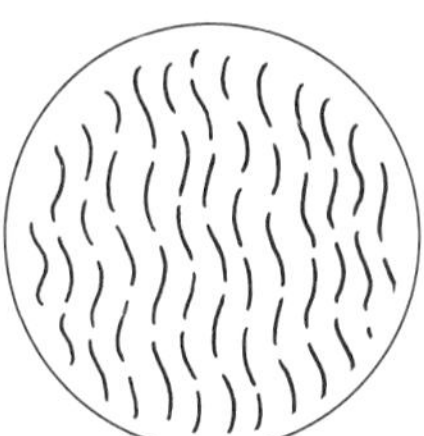

What Does Insulation Do?

Insulation is like a giant sleeping bag. It wraps the house in a layer of material that slows the rate at which heat is lost to the great outdoors.

Remember that heat always flows from warm to cold and it moves in three ways: by conduction, convection, or radiation.

Still air does not conduct heat well and is a relatively good insulator. However, in large spaces such as wall cavities heat can still be lost across the air space by convection and radiation. Insulation divides the air space into many small pockets of still air; this inhibits heat transfer by convection. At the same time, the insulation material reduces radiation across the space.

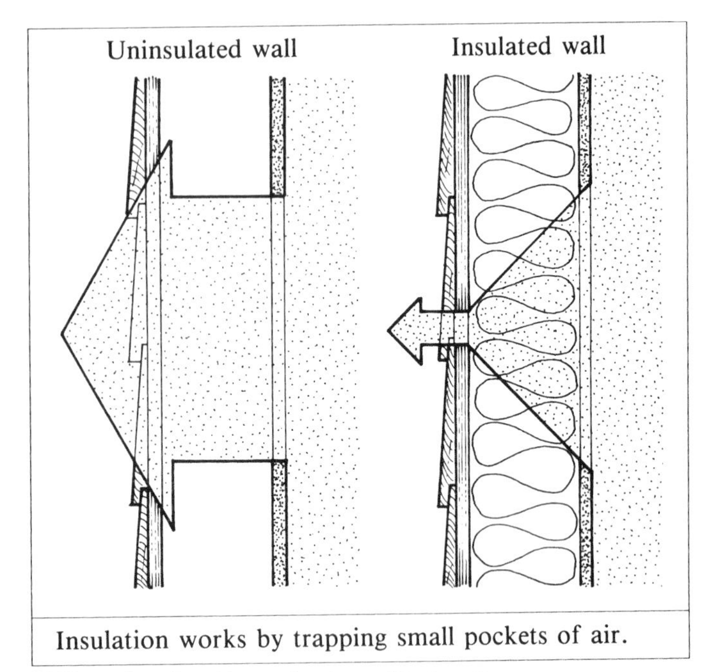

Insulation works by trapping small pockets of air.

What is the Insulation Value?

Years ago when the choice of insulation was limited, the measure of the effectiveness was its thickness. Products have changed and that rule of thumb can no longer be applied.

Insulation is now manufactured and sold by its **thermal resistance value** (called the "RSI" value) — a precise measurement of the insulation's resistance to heat flow. **The higher the resistance value, the slower the rate of heat transfer through the insulating material.**

One brand of insulation may be thicker or thinner than another, but if they both have the same RSI value, they will control heat flow equally well. Chapter 2 describes insulation materials and their RSI values.

Some insulation materials are marked with both "RSI" and "R" values. RSI values indicate thermal resistance in metric terms (Resistance System International), while R values represent the Imperial measurements. **Be careful not to confuse the two.**

For any insulation to work properly, it must be installed properly. Chapters 4, 5, and 6 describe how to install insulation in attics, basements and walls. But there are some common guidelines that apply wherever insulation is installed.

- The insulation must fill the space completely and evenly. Any blank spots or corners will allow convection to occur, sometimes allowing the heat to bypass the insulation completely.
- Wherever possible, avoid thermal bridges. As the name suggests, a thermal bridge is any solid material which directly connects the warm side of the envelope to the cold side, (e.g. a wall stud). Heat can easily escape, by conduction, along a thermal bridge. When insulation is installed on one side of the thermal bridge it acts like a road block, reducing heat flow.
- It's also important to install the appropriate thickness of insulation for the size of the space and when using loose fill, at the proper density.

Metric Conversion

to get	multiply	by
RSI overall	R overall	0.1761
RSI/mm	1 R/in.	0.00693

Nominal RSI	Nominal R
.7	4
.9	5
1.4	8
2.1	12
3.5	20
4.9	28
7.0	40
8.8	50
10.6	60

Measuring Up

Throughout this book we have used metric measurements and values, and given the Imperial equivalent in parentheses, e.g. RSI 3.5 (R20). We have used certain measurements as expressions, for example, a 38 mm x 89 mm stud is what we commonly know as a 2x4. In these cases we have chosen not to include the unit of measurement.

Imperial equivalents are included because most retrofit projects take place in houses that were built using this system of measurement.

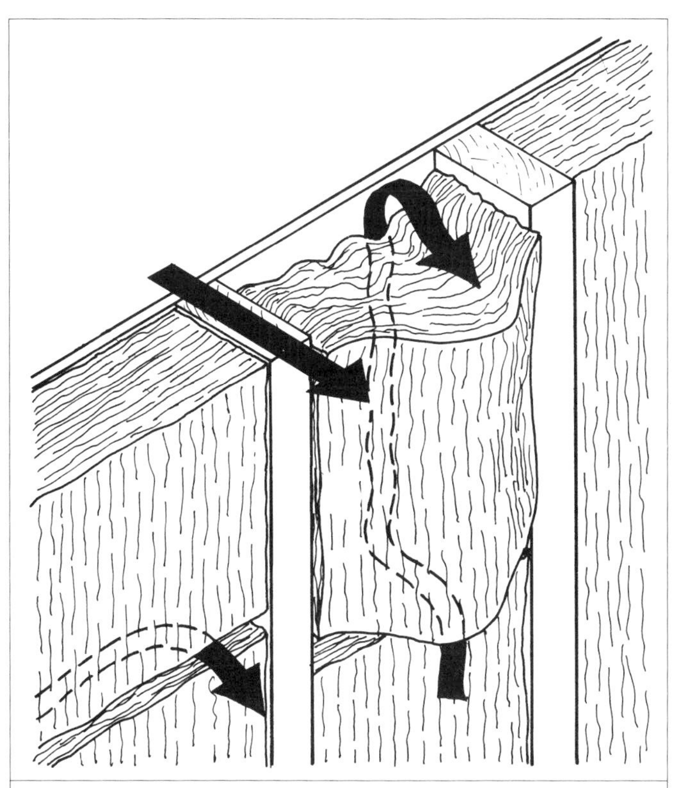

Wood studs provide a thermal bridge while gaps in the insulation allow convection currents.

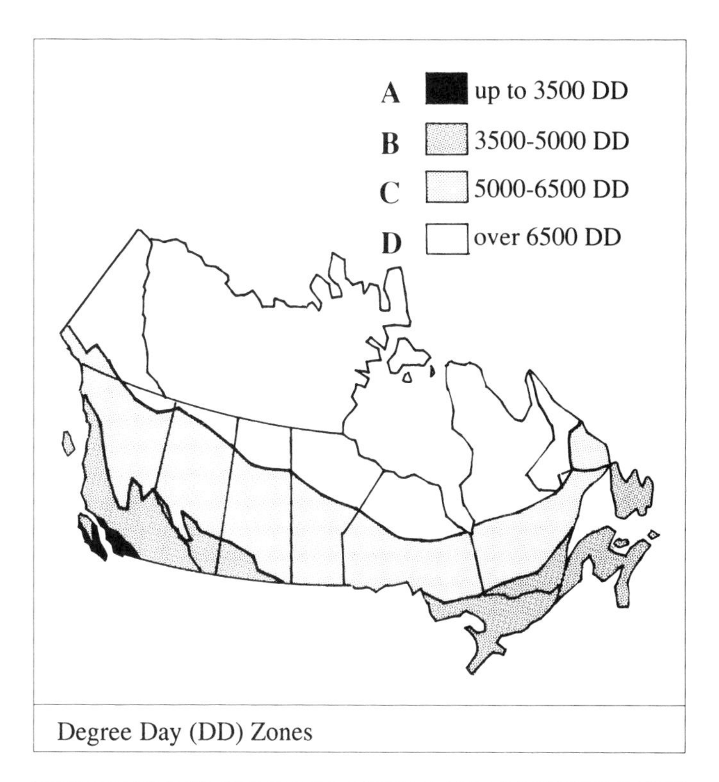

Degree Day (DD) Zones

Note: Each zone on the map represents an area which experiences a similar number of degree days. Degree days are a measure of heating demand, based on the difference between the average daily outdoor temperature and 18°C (65°F). Cumulative totals for the month or heating season are used to estimate heating energy needs.

How Much Insulation?

Your choice of how much insulation to add will depend on many factors.

- Local housing codes may specify minimum levels of insulation that should be added when upgrading.
- How much insulation is already in the house will partly determine how much you need to add.
- How the house is built will determine how much insulation may be practically added.
- Other work you're doing together with the reinsulating may make it practical to add higher levels of insulation.

The table on the right is a good guideline to follow to determine insulation levels for different areas of a house. To help you make your decision, you may wish to compare your plans with the new *Model National Energy Code for Houses* for your province or territory. The thermal resistance values listed in the Code are more accurate than the ones listed in the chart because they take into account the type of framing and heat loss through the framing materials. You can also contact the Energy Code Program for the recommended **minimum levels** of insulation that correspond to your method of construction, region and heating fuel type. Write to:

Zone		**A**	**B**	**C**	**D**
Walls	RSI	3.0	3.6	4.1	5.3
	R	17.0	20.0	23.0	30.0
Basement Walls	RSI	3.0	3.0	3.0	3.0
	R	17.0	17.0	17.0	17.0
Roof or Ceiling	RSI	4.5	5.6	6.7	9.0
	R	26.0	32.0	38.0	51.0
Floor (over unheated spaces)	RSI	4.7	4.7	6.7	9.0
	R	27.0	27.0	38.0	51.0

Natural Resources Canada
The National Energy Code for Houses
580 Booth Street, 18th floor
Ottawa, Ontario
K1A 0E4 Facsimile: (613) 943-1590

Part III Control of Air Flow

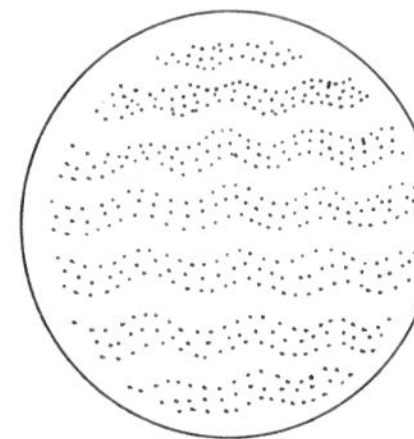

Why Control Air Flow?

The control of air flow provides many benefits.

- Greater energy and dollar savings.
- A more comfortable home without cold spots and drafts.
- Protection of the building materials from moisture damage.
- Improved comfort, health and safety; stale odors and stuffiness are eliminated and a safe supply of air for combustion appliances is assured.
- A cleaner and quieter home.

Controlling air flow involves three relatively simple activities:

- preventing uncontrolled air leakage through the envelope;
- providing for fresh air supply and the exhaust of stale air; and
- providing draft and combustion air for fuel-burning appliances.

The important point is that these activities must always be done together. Halfway measures won't do.

Air Leakage Control: Wind Barriers, Air Barriers, and Air Sealing

To be effective, insulation must trap still air. It must be protected from wind blowing through from the outside and from air escaping from the inside of the home.

The **wind barrier** is located on the outside of the envelope to protect the insulation from the circulation of outside air. Standard building materials such as exterior sheathing and building paper or new sheet materials such as spun-bonded olefin act as the exterior wind barrier.

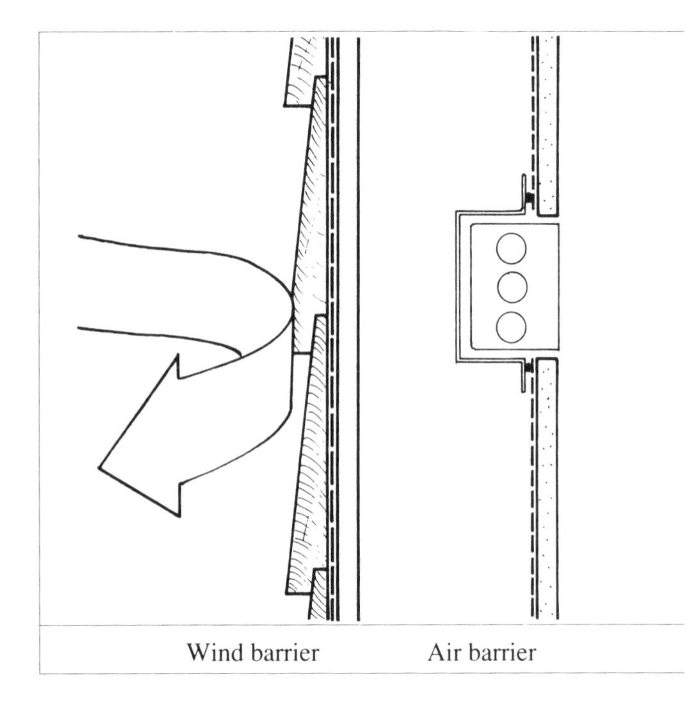

Wind barrier Air barrier

The **air barrier** blocks air flow from the inside to the outside. By doing this it serves two important functions.

- It reduces heat loss by preventing air from passing in and out through the envelope.
- It protects the insulation and structure from moisture damage caused when water vapor condenses in the envelope assembly.

The air barrier can be installed at any location in the envelope; it may even be combined with the wind barrier. Usually it's installed on the inside of the envelope, where it is kept warm. This protects the material from temperature extremes which can affect its durability. When installed on the warm side, the air barrier is often combined with the vapor barrier. (See section on vapor barriers under Part IV Control of Moisture Flow.) If located on the inside, the air barrier will also prevent convective heat loss when air circulates from the house into the wall space.

To be effective, the air barrier must be:

- resistant to air movement;
- rigid and strong enough to withstand air pressure differences;
- durable; and
- continuous, by sealing all seams, edges, gaps, holes or tears.

Because of the many components that make up the house envelope, including walls, foundations, doors and windows, it's impossible for any one material to surround the house completely and form the air barrier. The air barrier is actually a system made up of many different components that are sealed to each other. Typical components of the air barrier system are described below.

- Polyethylene, drywall, or plaster are used for large surfaces such as walls and ceilings.
- Windows, doors, hatches, vent dampers and any components that close an opening in the envelope also form part of the air barrier.
- In some cases, even structural parts of the building, such as the sill plate or rim joist, form part of the air barrier system.
- Caulking, gaskets and weatherstripping are used to seal the joints between the components to ensure that the air barrier system is continuous.

To make sure that you've accounted for a continuous air barrier, make a sketch of the wall, attic or foundation that you plan to retrofit. Then take a coloured pencil and trace a line through all the air barrier components on the sketch, without lifting the pencil from the paper.

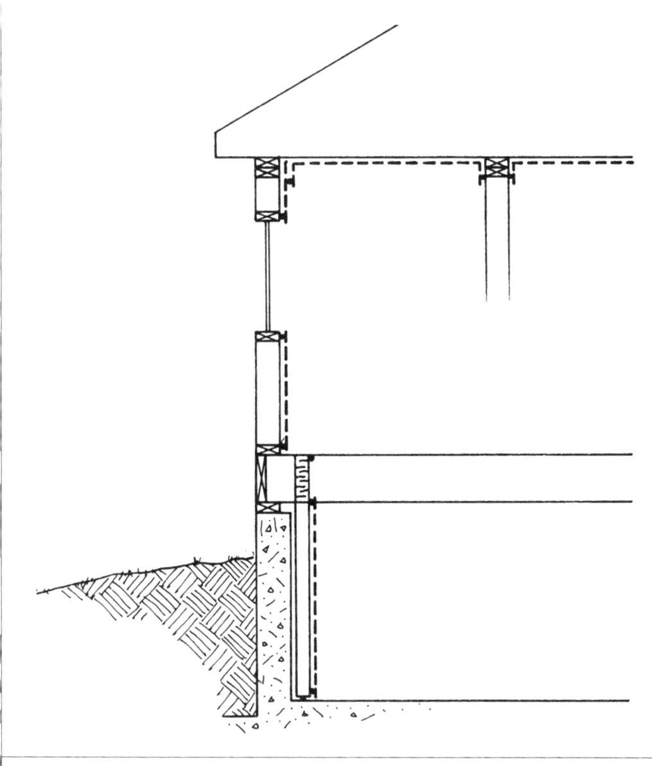

The air barrier is a system that joins many different building components.

How Tight an Air Barrier?

For an air barrier to work it must be as continuous and well sealed as possible.

But, if the air barrier is tight, how will fresh air get into the house? Firstly, most older houses are so loosely built that even after extensive air leakage control work enough air will still come in to provide ventilation. Secondly, remember that the air barrier is just the first step in the control of air flow. The other essential steps are providing ventilation air and air for combustion in a controlled manner. These steps may be necessary in houses where extensive retrofit and renovation work has been done, where the house is heated with a reduced flue-action heating system (e.g. electricity or high-efficiency gas) or where there are special ventilation needs. It is a good idea to take a systematic look at the moisture balance and ventilation needs of your house.

Providing Controlled Ventilation

Older homes were ventilated by opening windows and doors and by uncontrolled air movement, but this method was not always comfortable or effective. In cold, windy weather, too much air could be drawn into the house, causing high fuel bills and uncomfortable drafts. Often, in spring and fall, not enough fresh air would be supplied.

With uncontrolled air flow stopped by an air barrier system it's now possible to provide comfortable, effective ventilation year-round.

A system for controlled ventilation consists of four basic parts:

- a means of exhausting stale air and excess water vapor;
- a means of supplying fresh air;
- a way of distributing the fresh air throughout the house; and
- controls for operating the ventilation system.

Many homes already have parts of a complete ventilation system; it's just a matter of putting it all together and adding the missing components.

The **exhaust** function can be provided by kitchen and bathroom fans; these are already located conveniently in areas of high humidity. Driers should also be vented outdoors.

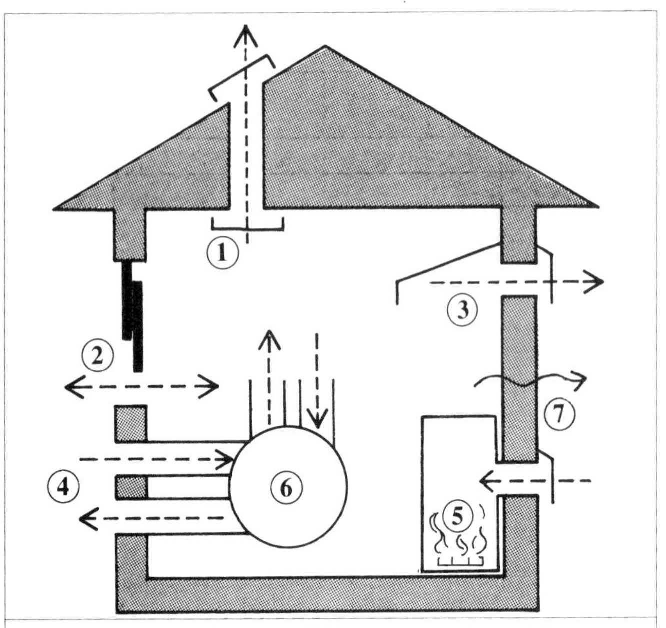

Houses can contain many different ways of providing ventilation: 1) bathroom fan, 2) open window, 3) kitchen fan, 4) exhaust or supply fan, 5) duct to return air plenum, 6) central ventilation system, 7) holes in the envelope.

Fresh air supply can be arranged for homes with a forced air heating system by installing an outside duct connected to the forced air system. Air is distributed by running the furnace fan on low speed, even when the furnace is not heating.

In homes with individual room heaters, fresh air can be provided by installing a central supply with a duct to each room and a fan to move the air. This system is easiest to install in bungalows. In colder regions the incoming air may need to be tempered or pre-heated.

Controls are usually connected to the exhaust side of the system; the supply side responds passively to replace the amount of air exhausted. One control method uses humidity as an indicator of how much air needs to be exhausted. There is usually an automatic setting on the exhaust fan for routine operation, with a manual override for cooking, showers or times when there are more people in the house.

How Much Ventilation?

Recognizing that homes are now built with tighter air barriers, the National Building Code now requires that new homes include a mechanical ventilation system with a minimum exhaust capacity of 1/3 an air change per hour (ACH). This means that one-third of the total volume of air in the home would be replaced by outside air every hour.

Not every home where retrofit work has been done will need this much mechanical ventilation capacity. Even after extensive air sealing work most older houses still allow considerable air flow through the envelope. Humidity levels and the appearance of condensation can provide a "rule of thumb" for judging ventilation needs. Generally, if there is only occasional or a small mist of condensation on double glazed windows on the coldest days then sufficient ventilation is present. Of course, houses where there is little moisture generated but there are pollutants due to hobbies, smokers, etc., will require more ventilation. The tightness of the air barrier system can be assessed using a fan test. (See Chapter 8 for more information on ventilation strategies.)

Providing Air for Combustion Appliances

A combustion appliance is any device that burns fuel; furnaces, fireplaces, woodstoves, gas stoves, gas water heaters and gas driers are all combustion appliances. Combustion appliances require air to burn the fuel and to provide draft for the chimney.

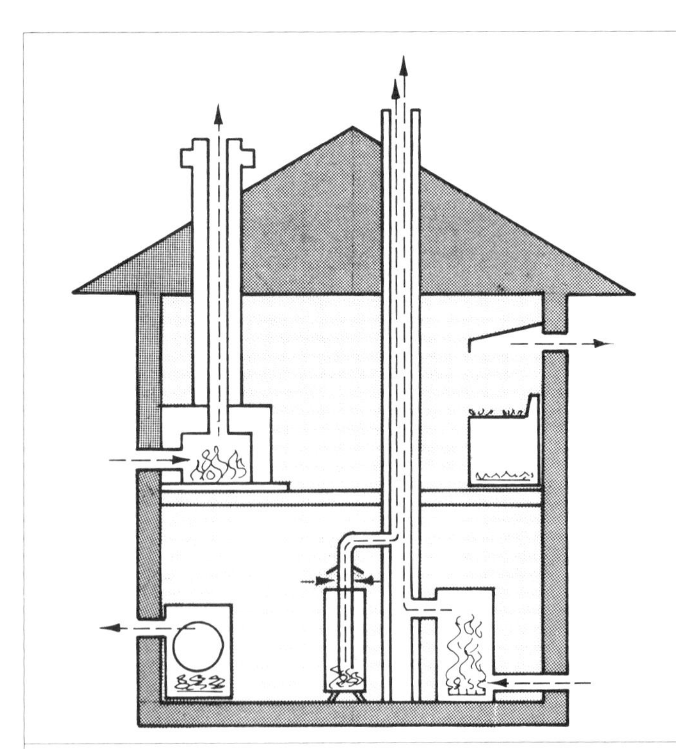

Fuel-burning appliances and exhausting appliances all need make-up air.

Older homes without a tight air barrier typically provide plenty of air for combustion and draft control through cracks and holes in the building envelope. Tighter houses, and houses where fans, exhaust appliances and fireplaces may be competing for air, can experience insufficient draft and even back-drafts. This can be a serious safety consideration where the draft in the chimney can be reversed and dangerous combustion gases can spill back into the house.

As part of any air leakage control program it's important to ensure that each combustion appliance has enough air to work properly. Chapter 8 provides more detail on combustion air strategies.

Part IV Control of Moisture Flow

Why Control Moisture Flow?

Control of moisture in all its forms is important in order to make our homes durable and comfortable. Building practices such as flashing, roofing, and basement damp-proofing successfully protect the home from liquid water.

It is equally important to control the movement of water vapor, providing added protection for the house structure and helping to maintain indoor humidity at a comfortable level.

Controlling moisture involves three strategies:
- construction techniques which keep moisture away from the structure;
- producing less moisture; and
- exhausting excess moisture.

Sources of Moisture in the Home

Even some houses that are apparently dry, with no leaks in the basement or roof, can have moisture problems. Where does all the moisture come from? There are a number of major sources that are not always obvious:
- occupants and their activities;
- wind-blown rain in walls;
- damp basements; and
- moisture that is stored in building materials and furnishings.

An average family of four will generate about 63 litres of water a week through normal household activities. Where basement damp-proofing is inadequate, groundwater in the soil can migrate through the foundation by capillary action and evaporate on the surface of the wall or floor. Finally, during damp, humid weather the building materials and furnishings will absorb moisture from the air, then expel it during the heating season.

Quantity of Moisture Added to the Air Through Various Household Activities

Activity (for a family of four)	Moisture litres per week
Cooking (3 meals daily for 1 week)	6.3
Dishwashing (3 times daily for 1 week)	3.2
Bathing (0.2 litres per shower) (0.05 litres per bath)	2.4
Clothes washing (per week)	1.8
Clothes drying indoors, or using an unvented dryer Per week	10.0
Floor mopping per 9.3 m² (100 ft.²)	1.3
Normal respiration and skin evaporation from occupants	38.0
Total moisture production per week	63.0

Despite all this water produced each day, most older houses have "dry" air in winter to the point where they have to install humidifiers. Why?

Cold outdoor air cannot carry much water vapor. In older homes, uncontrolled air flow brings colder, drier air indoors and forces the warm, moist household air out through openings in the upper walls and attic. The air quickly escapes through the uninsulated envelope without cooling down enough to cause condensation.

When insulation is added, the building exterior becomes much colder. Unless additional protection is provided, **water can condense in the building structure.**

How? Remember that cold air can hold much less moisture than warm air. As the warm moist air cools in the cold outer layers of the building, the water vapor it holds may condense as liquid or, if it's cold enough, as frost. This can reduce the effectiveness of insulation and even cause rot, paint peeling, buckled siding, mould growth, and other problems.

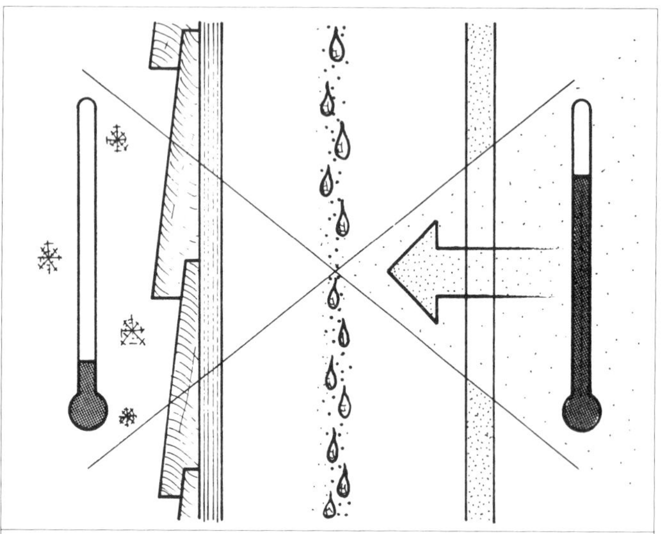

Water vapor condenses as either liquid water or frost when it reaches the dew point.

How Much Humidity?

Humidity levels above 20 per cent help prevent dry, sore throats and should make the air feel warmer and more comfortable. Moist air will also eliminate static electricity in the house and help to protect plants and preserve your furniture.

On the other hand, humidity levels over 40 per cent can cause frosting and fogging of windows, staining of walls and ceilings, peeling paint, mould growth and odors. When relative humidity is over 50 per cent, airborne diseases become more difficult to control.

Condensation on your windows can provide a good indication of the relative humidity. You may, however, want to install a humidity sensor or humidistat to keep more accurate measurements on the humidity levels.

Keeping the Structure Dry

Four strategies are used to keep the structure dry.

1) Exterior weather and moisture protection requires building paper, siding, flashing, gutters and other construction techniques to shed water and repel wind-driven rain. It also involves below-grade measures like proper drainage, grade slope and damp-proofing to protect the foundation from groundwater leaks or from moisture movement by capillary action.

The building envelope must shed water from the roof to the footings.

2) Reducing moisture at the source involves producing less moisture in the first place and exhausting moist air and bringing in drier air. Detailed strategies are outlined in Chapter 8.

3) Preventing moist indoor air from getting into the envelope requires a vapor barrier to reduce moisture movement by diffusion and an air barrier to prevent moisture movement by air leakage.

Air barriers are described in Part III, Control of Air Flow. Although less moisture can be moved into the envelope by vapor diffusion, it is still important to provide a vapor barrier. An effective vapor barrier is:

- resistant to vapor diffusion;
- durable; and
- installed on the warm side of the insulation; but
- it needn't be continuous.

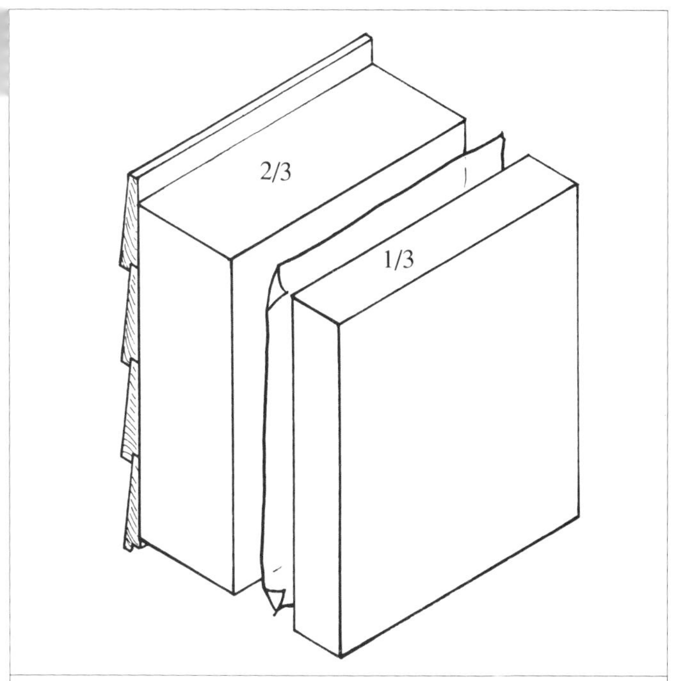

Up to 1/3 of the insulating value can be installed on the warm side of the vapor barrier.

A number of building materials resist vapor diffusion well enough to be used as vapor barriers. These include polyethylene, oil base paints and special vapor barrier paints, some insulation materials and exterior grade plywood. Different materials may act as the vapor barrier in different parts of the house.

The same material may work as both an air barrier and a vapor barrier, provided it meets both requirements and is properly installed. Polyethylene sheets and foil-backed gypsum drywall can both combine these functions. To avoid confusion of terms, when a material is doing both jobs it will be called an *air and vapor barrier*.

As a general rule, the vapor barrier should be on the warm side of the insulation. In some cases, however, the vapor barrier can be located within the wall or ceiling assembly, provided that at least two thirds of the insulation value of the wall is on the cold side of the vapor barrier. This ratio should be adjusted for houses with high interior humidity or for homes in extremely cold climates.

4) Letting the envelope "breathe" to the outside allows the house to deal with seasonal fluctuations in humidity, and to release any moisture that does penetrate the envelope from the interior or exterior. This is accomplished in two ways. The materials of the envelope are layered, with those most resistant to vapor diffusion located on the warm side of the envelope, and the least resistant (such as building paper) located on the outside. In this way, any vapor which penetrates the envelope can escape to the outside.

Some wall systems work well with a relatively impermeable insulated sheathing because the interior wall cavity temperatures are kept high. As a precaution, when retrofitting a wall, always ensure that the interior surfaces are vapor resistant.

Some siding applications have an air space immediately behind the exterior finish to promote drying out of materials that have been soaked by rain or dampness. This air space also provides an escape route for any moisture that has penetrated the wall cavity from the indoors. This type of installation should **not** be used with insulated sidings as convection in the air space will negate the effect of the insulated backer board on the siding.

Part V The House as a System in Action

The examples given in this chapter highlight the need to consider the House as a System when planning any retrofit work. Pay special attention to the air and moisture balance and the effect of the retrofit on the heating system and ventilation.

For major retrofit projects you may have to anticipate the need for changes to the heating system or house ventilation and include these changes in the work plan. When doing smaller projects which are spread out over time, monitor your house carefully after each project to assess the impact of the changes. At some point adjustments to the heating system and house ventilation will likely be essential to keep the system working properly.

Further information on the house as a system is included throughout this book. Later chapters which describe the how-to of insulating your home will also include recommended methods for controlling air and moisture in walls, basements and attics. The chapter on Comprehensive Air Sealing includes strategies for moisture control and the chapter on Operating Your House describes strategies for providing ventilation and combustion air.

Part VI Older Homes and Heritage Buildings

Whether your home is 50, 75 or 100 years old, it represents a part of our architectural heritage. Older homes deserve special consideration when retrofitting. Maintaining the durability of the structure is especially important. Homes over 50 years old may incorporate unusual construction details and materials that make it necessary to improvise and adapt standard retrofit methods. Retrofit work will need to be done with sensitivity to the design, materials and special features of the home. The retrofit will need to minimize changes to the building's appearance and emphasize repair, rather than replacement of building components.

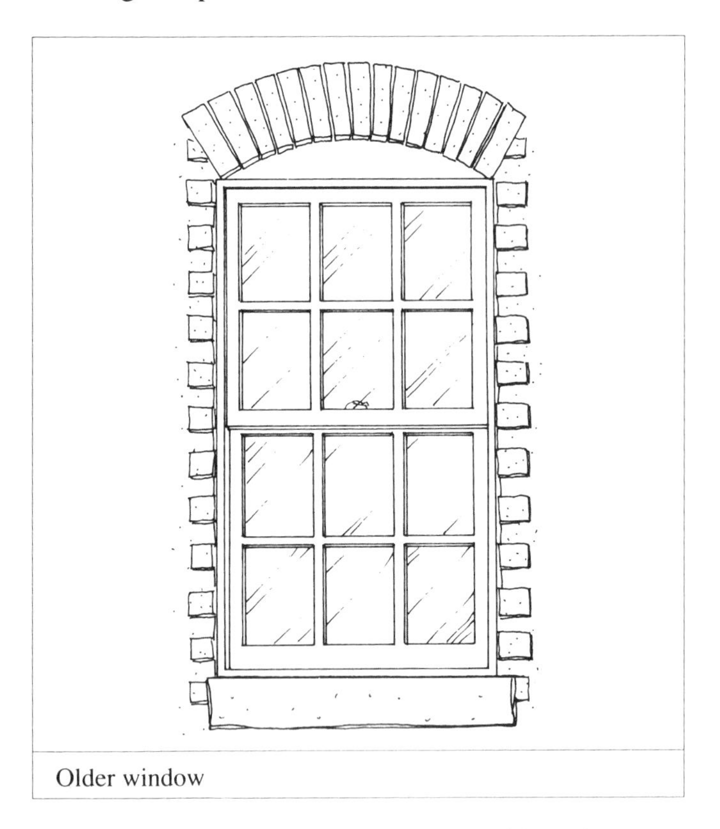

Older window

While there are bound to be some sacrifices in energy efficiency, with a little more planning and care you can do a lot to make older homes more comfortable, durable and energy efficient. Extra care at the planning stage involves assessing the home from several aspects.

- Heritage: what features have to be preserved? What later changes, (if any) should be removed?
- Maintenance and repair: what areas need attention? Do any needed repairs indicate moisture or structural problems that should be corrected?
- Energy efficiency: what improvements can be made while maintaining the heritage quality of the building?

Some sample guidelines and opportunities for retrofit of older homes are given below.

Air Sealing: Comprehensive air sealing is one of the least obvious and most effective retrofit projects for older homes.

Heating System: A total tune-up of the heating system is another inexpensive, effective and invisible measure for older homes.

Insulation: Preserving the structure is especially important; take extra care to provide a vapor barrier and air barrier when insulating. Basements and attics can often be reinsulated without affecting appearance. Where it's desirable to preserve both the interior and exterior wall finishes, blowing insulation into the cavity of a wood frame wall is an option. Often, older homes have had the original exterior finish replaced with a more 'modern' but less appropriate material. Insul-brick may have replaced the original stucco, or 'perma stone' the original cement parging. These situations provide the opportunity to retrofit from the exterior and, at the same time, copy the original finish.

Windows: Windows are one of the most important aspects of a home's originality. Careful weather-stripping of older, single-pane, wood frame windows will do much to improve their energy efficiency. If the original wooden storm windows have been destroyed, it's possible to have custom wood storms made to order. If the object is to preserve the appearance of the building, avoid metal storms or storm and screen combinations.

If exterior wood storms are not desirable because of the maintenance factor, interior storms offer a good alternative. These are less noticeable than exterior metal storms and they can be made to fit on the sash or the window trim. If the window sash is badly deteriorated, replacement units can be made to fit the existing frame.

Doors: Preserving the original doors is important to the overall appearance of an older home. Careful weatherstripping will improve their performance. As with windows, avoid aluminum storms. A better alternative is to restore the enclosed vestibule that is found in most older homes.

2 Materials

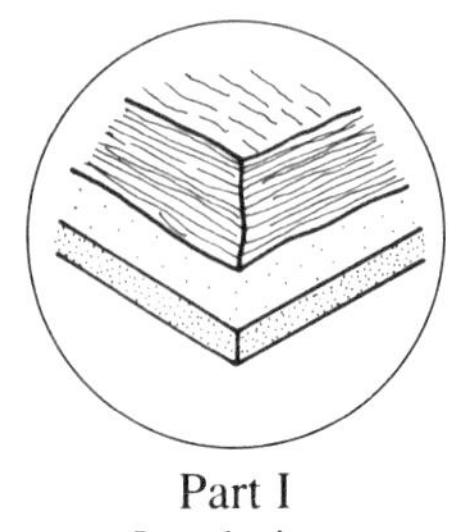

Part I
Insulation

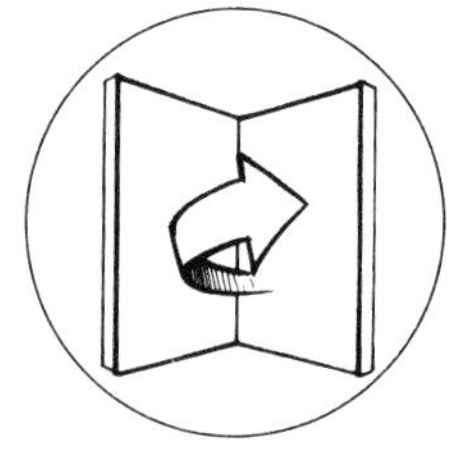

Part II
Air Barrier Materials

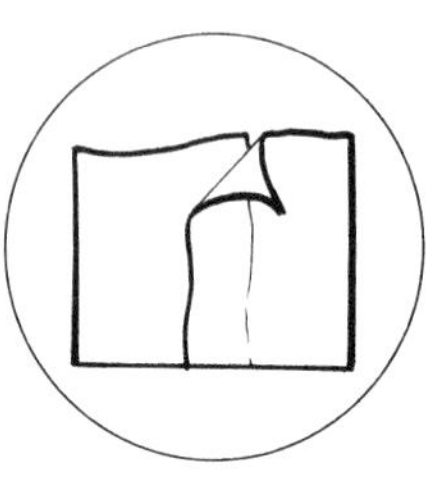

Part IV
Health and Safety
Considerations

Part III
Vapor Barrier Materials

Introduction

Whether you're doing the work yourself, or hiring a contractor, it's important to know which are the right materials for your particular job. Choosing the right materials and installing them properly will ensure that the finished product lives up to your expectations.

This chapter will describe the three types of materials used when *Keeping The Heat In*: insulation, air barrier materials and vapor barrier materials.

Review the How To chapters to identify those types of insulation and materials most suited to your project, and then study your options in detail in this chapter. **Use the information to choose the most appropriate material.**

Part I Insulation

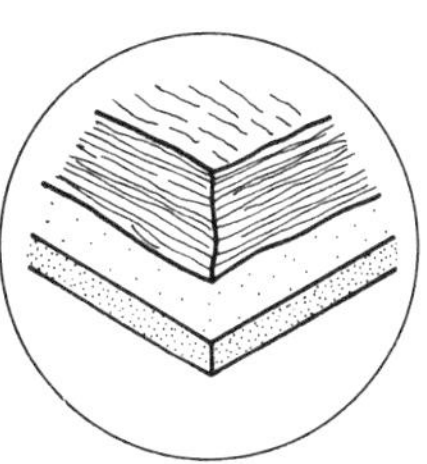

Requirements for Insulation

Insulation is the one material whose main function is to keep the heat in. In order to be effective, insulation must be:

- resistant to heat flow;
- able to fill the space completely and evenly,
- durable; and,
- for some locations, able to withstand exposure to heat or moisture.

Several different insulation materials may be used at different locations in the house envelope, depending on the space available for the insulation, ease of access and other installation requirements.

The Proper Choice of Insulation

The proper choice of insulation depends on its final use. In most applications, good resistance to heat flow is not the only thing you will have to consider. In **specific** situations, the insulation may also need to have any number of the following properties.

- Is the insulation able to resist high temperatures?
- Is it resistant to moisture flow; can it reduce the movement of water vapor?
- Does the insulation resist air movement; can it act an an air barrier?
- When used as proposed, does the insulation need a fire-rated protective covering?

Once you have matched the material properties with the specific application, consider installation factors.

- Is it relatively easy to install?
- Is it the best buy for the space available (either high insulating value per dollar if you have lots of open space, or high insulating value per thickness if space is restricted)?
- Is it locally available?
- Will it be easy to install the insulation to fill the space completely?
- Can it conform to surface irregularities?
- Is it rigid enough to provide support for finished materials or resist pressures against its surfaces?
- Does one insulation require more accessory products than another (fire protection, framing, air and vapor barrier)?

In short, the choice of insulation will largely depend on how it will be used. The How-To chapters on insulating walls, basements and attics discuss the types of insulation commonly used for these applications. Fortunately, particular insulation jobs will quickly eliminate some materials, making the choice much easier.

Naturally, cost is a factor in choice of material. Generally the cost per RSI value is lower for loose fill or batt-type materials than for rigid board or foam-type insulations. However, the price of the basic material is just one aspect. In some cases, high material costs may be offset by lower installation costs or the preference of the installer for a particular insulation technique. A better comparison can be made using the installed cost. This would include the cost of the insulation material plus the cost of required accessories and installation.

Insulations are manufactured from a wide range of materials, including melted glass spun into fibres, expanded volcanic rock, recycled newsprint, and foam plastic.

However, there are only four basic forms of insulation, which provide a ready means of classification: **batt or blankets, loose fill, rigid or semi-rigid boards, and foamed-in-place**. All of these categories are described in detail on the following pages.

Note: Unless otherwise indicated, all the insulation materials listed require the presence of an air barrier and a vapor barrier.

Insulation Summary

Batt or Blanket Insulation

Batt or blanket insulation is relatively easy to install in accessible spaces such as exposed wall cavities and some attics. It conforms to slight surface irregularities and can be cut to fit. Safety equipment and protective clothing are required during installation.

Batt Insulation

Mineral Fibre

- 0.022 RSI/mm (3.2 R/in.) glass fibre
- 0.023 RSI/mm (3.3 R/in.) mineral wool
- Available in batts or continuous rolls (blankets).
- Will not settle.
- Some products are non-combustible. (Check with the manufacturer.)

Loose Fill Insulation

Loose fill insulations are made from a variety of materials, with particles ranging in texture from granular to fluffy. Loose fill is excellent for filling irregular or inaccessible spaces. It is suitable for walls and floors and excellent in attics and enclosed spaces such as roofs where the space between the joists may be irregular or cluttered with obstacles. It is often handy for filling small spaces or covering ceiling joists. It is not appropriate for below-grade application.

Loose fill insulation may be blown or poured. Pouring insulation will generally require more material than blowing insulation to achieve a specified RSI value. It is useful for topping-up existing insulation in attics and accessible enclosed wall cavities and filling in cracks and uneven spaces.

Proper installation of blown loose fill insulation usually requires an experienced, well trained technician. To achieve the full RSI value the material must be installed following the manufacturer's instructions.

Safety equipment and protective clothing are required during installation. **The most important aspect of installation is following the manufacturer's instructions.**

Blown-In Insulation

Cellulose Fibre

- Cellulose fibre insulation is made from shredded newsprint treated with chemicals that resist fire and fungal growth and inhibit corrosion.
- Because of its small particle size, cellulose fibre is able to fill around obstructions such as nails or electrical wires within cavities.
- Cellulose may reduce air leakage if installed to the proper densities.

Blown

- Average RSI of 0.025/mm (3.6 R/in.), depending on the paper and chemical mix and the blown density.
- **The proper blown density for enclosed cavities is 56 to 72 kg/m^3 (3.5 to 4.5 lb./ft.^{3}).**

Poured

- Insulating value of 0.024 RSI/mm (3.4 R/in.)
- Follow the manufacturer's instructions for the proper application techniques.

Glass Fibre

- Loose fill glass fibre is a similar material to glass fibre batts, but is chopped up for blowing or pouring applications.
- Hand-poured glass fibre works best in open horizontal surfaces such as attics. Blown glass fibre can be used in both horizontal and vertical applications, but may be difficult to install in cavities that are partially blocked by nails, framing, electrical wiring etc.
- For walls, the density of application is usually 2 to 2 1/2 times the recommended rate of application for horizontal surfaces.
- Some glass fibre insulation is classified as non-combustible. Check the manufacturer's specifications.

Blown

- It has an average insulating value of 0.020 RSI/mm (2.9 R/in.) depending on density.
- Follow the manufacturer's instructions for the proper application techniques.

Poured

- Its insulating value is 0.021 RSI/mm (3.0 R/in.).
- Follow the manufacturer's instructions for the proper application techniques.

Loose Fill Insulation

Mineral Wool (Slag and Rock Wool)

- This material is treated with oil and binders to suppress dust and and keep its shape; a lubricant is added for blowing purposes. It is similar to glass fibre in appearance and texture.
- Mineral wool is suitable for accessible attics and inaccessible areas such as wood-frame roofs, walls, and floors.
- For walls, the density of application is usually 2 to 2 1/2 times the recommended rate of application for horizontal surfaces.
- It is a good material for insulating around masonry chimneys, as it will not support combustion.

Blown

- Mineral wool has an average insulating value of 0.021 RSI/mm (3.0 R/in.), depending on blown-in density.
- Follow the manufacturer's instructions for the proper application techniques.

Poured

- Its insulating value is 0.022 RSI/mm (3.2 R/in.).
- Follow the manufacturer's instructions for the proper application techniques.

Vermiculite

- Vermiculite is an expanded mica material.
- There are two types of vermiculite—untreated and treated. Untreated vermiculite absorbs moisture. Treated (water-repellant) vermiculite is coated with asphalt for use in areas of high moisture.
- Untreated vermiculite has an average insulating value of 0.016 RSI/mm (2.3 R/in.); treated vermiculite has an average insulating value of 0.017 RSI/mm (2.5 R/in.).
- Usually hand-installed.
- Suitable for both horizontal and vertical applications.
- For vertical applications, the material is poured into the wall cavity and packed down with a heavy weight to make sure that the cavity is filled and to prevent future settling.
- Follow the manufacturer's instructions for the proper application techniques.
- Due to its high density, this is not the best material to use where high R-values are desired.

Rigid Board Insulation

Board insulations are manufactured from glass fibre or foam plastic materials. These materials have a high insulating value per unit thickness although the cost per RSI value is greater than for loose fill or batt/blanket insulations.

Insulating boards are lightweight and easy to cut and handle. Fitting them into irregular spaces, however, can be a tedious process. Some boards are now available with an attached fire-resistant, moisture-resistant, or decorative covering. It is also possible to purchase specially designed boards that come with their own system of attachment. Regular board materials can be ordered pre-cut to specific sizes for an additional cost.

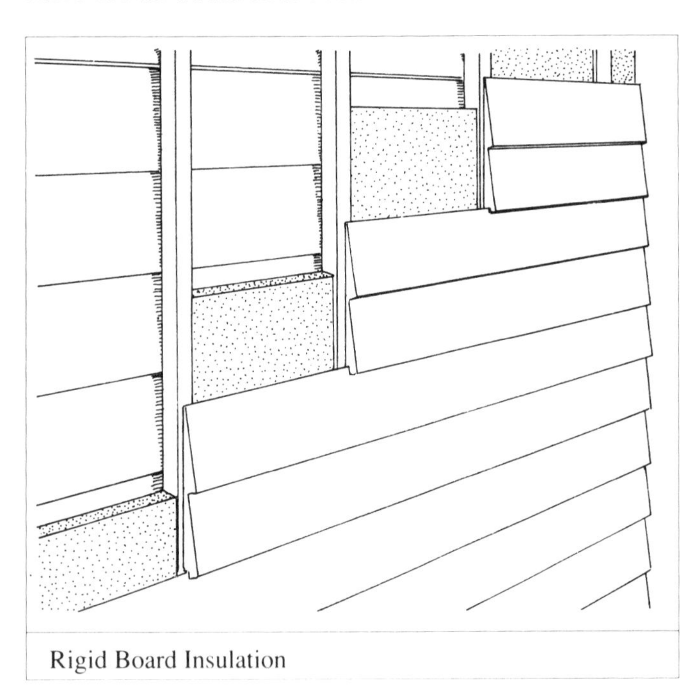

Rigid Board Insulation

Glass Fibre Boards

- There are two types of high density, semi-rigid glass fibre board commonly used in residential applications. One is designed specifically for below-grade exterior use, the other is an above-grade exterior sheathing.
- The above-grade type comes covered with a water-repellent breather-type building paper.
- The below grade type has an insulating value of 0.029 RSI/mm (4.2 R/in.).
- The exterior sheathing has an insulating value of 0.031 RSI/mm (4.4 R/in.).

Expanded Polystyrene

- Expanded polystyrene is produced by bonding coarse beads into rigid "foam plastic" boards. It is often referred to as "bead board".
- Expanded polystyrene is manufactured in two densities:
  - Low density, giving 0.026 RSI/mm (3.7 R/in.)
  - High density, giving 0.028 RSI/mm (4.0 R/in.)
- The high density board is more resistant to moisture than the low density board and can be used on the exterior of foundation walls in dry, sandy soils.
- Polystyrene must be protected from prolonged exposure to sunlight, solvents and some sealants. Use compatible sealants only. Ask your dealer fo information.
- Expanded polystyrene requires covering with a fire-resistant material.

Extruded Polystyrene

- Extruded polystyrene is a foam plastic board wit fine, closed cells containing a mixture of air and refrigerant gases (fluorocarbons).
- Extruded polystyrene is manufactured in two densities:
  - low 0.033 RSI/mm (4.7 R/in.) to 0.035 RSI/mm (5.0 R/in.).
  - high 0.035 RSI/mm (5.0 R/in.).
- Polystyrene must be protected from prolonged exposure to sunlight or solvents.
- If joints are sealed properly, extruded polystyrene can perform as an air barrier and certain thicknesses may perform as a vapor barrier.
- When installed on the interior, it must be covered with a fire-resistant material mechanically fastened to the building structure.

Polyurethane and Polyisocyanurate Boards

- These plastic boards are made of closed cells containing refrigerant gases (fluorocarbons) instead of air.
- Boards usually come double-faced with foil, or sometimes bonded with an interior or exterior finishing material.
- Faced boards have a typical insulating value of 0.040 RSI/mm (5.8 R/in.) to 0.050 RSI/mm (7.2 R/in.) and come in a variety of sizes.
- Must be protected from prolonged exposure to sunlight and water.
- Must be covered with a fire-resistant material.
- Can act as an air barrier (if seams are well sealed) as well as a vapor barrier.
- Use is generally limited to areas where a high RSI is desired and space is at a premium.

Phenolic Foam Boards

- Phenolic foams are manufactured from phenol formaldehyde resin. Some panels have a water repellant exterior skin on both sides.
- Suitable for areas where space is at a premium but high RSI values are required.
- **Typical insulating value of 0.030 RSI/mm (4.3 R/in.) open cell, and 0.058 RSI/mm (8.3 R/in.) closed cell (based on manufacturer's literature).**
- Phenolic foam must be protected from exposure to sunlight and water.

Spray-Foam Insulation

This type of insulation is mixed on the job site by the contractor/installer. The liquid foam is sprayed directly onto the building surface or poured into enclosed cavities using a spray-gun driven by a pump. The foam expands in place and sets in seconds. The installation contractor should be trained in the application of the specific product.

Polyurethane Foam

- Polyurethane is a pale yellow foam of closed cells containing refrigerant gases (fluorocarbons).
- It has a typical insulating value of 0.042 RSI/mm (6.0 R/in.). Other values are sometimes quoted, but this value takes into account the loss of refrigerant gasses over time.
- The foam is sprayed onto surfaces in layers less than 50 mm (2 in.) thick and hardens in seconds.
- The foam can expand up to 28 times its original size and should not be used in enclosed cavities.
- Polyurethane foam can be used as an air barrier but not as a vapor barrier.
- It must be protected from prolonged exposure to sunlight and requires covering with a fire-resistant material when used indoors.

Semi-Flexible Isocyanurate Plastic Foam

- A combination of isocyanurate, resins and catalysts form this open-celled semi-flexible plastic foam insulation.
- **Manufacturer's literature lists its insulating value as 0.030 RSI/mm (4.3 R/in.).**
- There are some limitations on the thickness that can be applied.
- This material can be used as an air barrier.
- It requires covering with a fire-resistant material.

Part II Air Barrier Materials

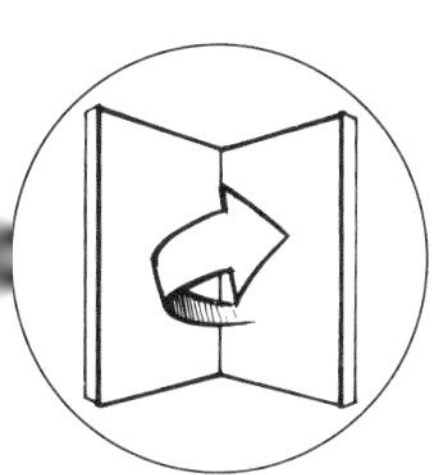

Requirements for Air Barrier Systems

The air barrier system is an important part of any retrofit job. It is the main means of protecting the building structure and the insulation from moisture damage. In order to be effective, the air barrier system must be:

- resistant to air movement;
- continuous, completely surrounding the envelope of the house;
- strong enough to withstand wind pressures; and
- durable.

A variety of materials are used throughout the envelope to act as the air barrier. In some cases, building materials, such as drywall, baseboards or structural members are incorporated into the air barrier by sealing them to adjoining materials.

Air Barrier System Components

The most common components of an air barrier system are:

- sheet or rigid materials for large surfaces;
- caulking and gaskets for joints between materials that don't move; and
- weatherstripping for joints that do move.

Choosing Air Barrier Materials

If the material offers resistance to air flow, strength and durability, consider these installation factors:

- Is it easy to install?
- If installed in a concealed location will it last the life of the building *or* will it be accessible and easily repaired?
- Is it compatible with other materials in the system? Can it be successfully sealed to adjacent materials?
- Is the choice of material appropriate to the other work being done on the home? Some renovation work will permit the installation of a new sheet material air barrier, while other jobs may require comprehensive air sealing work instead.
- Does it provide other functions such as insulation or vapor barrier?

Sheet Materials

Polyethylene sheeting

- Available in wide sheets, minimizing the number of seams required.
- Seams and edges should be supported on both sides to maintain the seal.
- 0.15 mm (6 mil) thickness is now commonly installed as an air barrier because it is more durable on the construction site.
- Polyethylene should be protected from exposure to sunlight. When it will be exposed to sunlight over extended periods (when wrapping the exterior of a house for example), a U.V. stabilized polyethylene should be used.
- Materials should be clear, made from virgin material, and labeled. It should conform to the Canadian General Standards Board (CGSB) standard for polyethylene.
- Polyethylene can also function as a vapor barrier.

Spunbonded olefin

- Available in wide sheets, minimizing the number of seams required.
- Acts solely as an air barrier, does not function as a vapor barrier.
- Generally used to wrap the exterior of a house; often bonded to exterior glass fibre sheathing .
- When installed on the exterior it acts as a wind barrier, preventing wind from reducing the effective RSI value of insulation.
- Olefin should be protected from extended exposure to sunlight.

Rigid Materials

Most solid building components will act as barriers to air. These components include drywall, plaster, plywood, glass, wood, and poured concrete (not concrete blocks). Insulating materials such as rigid foam boards also act as air barriers. To be effective, however, the seams between these various materials must be sealed with the appropriate caulking, weatherstripping or gasket. For example, caulking can be used between a baseboard and wall and baseboard and floor, linking the air sealing qualities of three different building components. The combination of rigid air barrier materials forms the house's air barrier system as long as the joints are well sealed.

Sealants

Caulking is used to seal joints between building components. Most joints move because of changes in moisture and temperature in the building. Some caulking materials can seal a larger joint and accommodate more joint movement than others. **Make sure that the caulking you use is compatible with the surfaces you are applying it to!**

Caulkings are not permanent and will have to be maintained. They also vary in their durability, compatibility with other materials, paintability and curing time. All sealants will require some extra ventilation of the house after application to let the material cure. Typical curing time will be no more than 2 or 3 days for interior application.

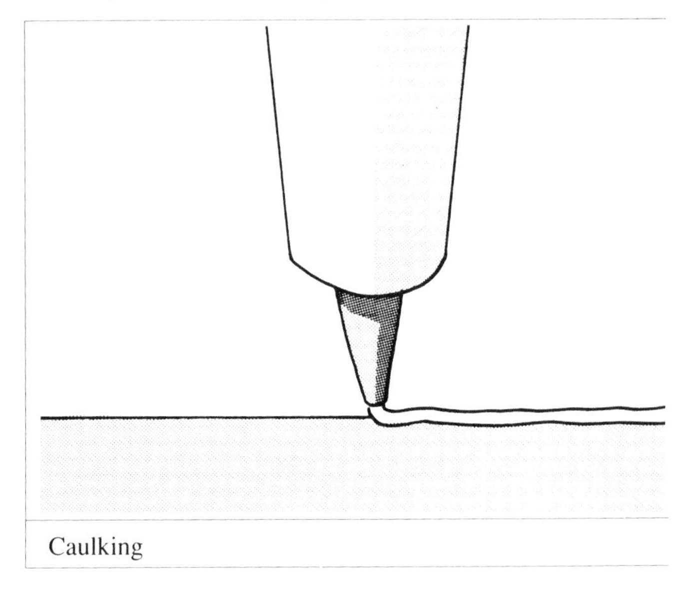

Caulking

Caution: Exterior products should not be used indoors – they may be hazardous when their volatiles are inhaled in a confined space over an extended period of time. Check manufacturer's literature very carefully.

Acoustical sealant

- Bonds to most surfaces, but excellent for use on metal, concrete and gypsum board.
- Excellent for sealing the joints in polyethylene air and vapor barriers, but should be used only where it is sandwiched between two materials.
- Mechanical support (e.g. staples) is required when used to seal the joints in polyethylene.
- Maximum joint width of 16 mm (5/8 in.); accepts some joint movement (10%).
- Very durable (20 year life expectancy).
- Non-hardening and not paintable; use should be limited to unexposed applications.

Acrylic latex
- A water-based emulsion sealant.
- Excellent for non-porous surfaces such as aluminum, glass and ceramic tile, but may be used to seal joints in wood surfaces.
- Maximum joint width of 9 mm (3/8 in.); limit use to joints where little to no movement is expected.
- Durable (10 year life expectancy).
- Available in a wide range of colours; paintable.
- Can attract dust over time, especially when used next to floors.

Butyl rubber
- A synthetic rubber sealant.
- Bonds to most surfaces; particularly suited to metal and masonry.
- Appropriate for a joint width up to 12mm (1/2 in.) accepts some movement of joint (5-10%).
- Durable (10 year life expectancy).
- Available in a variety of colours; paintable after one week's curing.
- Ventilation required during application and curing (up to 3 days).

Silicone sealant
- Solvent-free silicone compound; produces a flexible water-tight seal upon curing.
- Good adhesion to most surfaces; primers may be required on wood, steel or anodized aluminum.
- Excellent for large moving joints: up to 25 mm (1 in.) joint width and 12-50% joint movement.
- Highly durable (over 20 year life expectancy).
- Most types are not paintable.
- Available in several colours and clear; clear silicone is particularly suited for sealing highly visible joints where the caulking should not be noticeable.
- Ventilation is required during application and curing.

Polysulfide sealant
- Produces a flexible sealant upon curing.
- Ideally suited for use on stone, masonry, and concrete surfaces when used with a special primer.
- Maximum joint width of 25 mm (1 in.); will accept joint movement of 12-25%.
- Excellent durability (over 25 years life expectancy).
- Available in several colours; paintable.
- Ventilation is required to remove potentially toxic vapors.

Urethane foam sealant
- Available in a dispensing system with spray nozzles or individual aerosol spray cans.
- Foam types are available with different rates of expansion depending on ingredients and the amount of pre-curing. Check the cans carefully for details on sizes of cracks that can be filled. Some types expand slowly and moderately while others expand quickly and greatly. Use gloves and a drop cloth.
- Bonds well to most surfaces except polyethylene, teflon, or silicone plastics.
- Very good for filling larger joints and cavities where conventional sealant materials would not be suitable: header/joist intersections, around plumbing and vent openings.
- Urethane foam should not be used at window headers since it can transfer structural loads if wall settles.
- Good durability (10-20 years).
- Like all insulating foams it must be covered with a fire-resistant material.
- Ventilation is required to remove potentially toxic vapors.

Stove or muffler cements
- For use in areas where high temperatures are experienced, but where there is no joint movement.
- Typical application – used in conjunction with other materials for sealing around masonry or factory built chimneys.

Gaskets

Several specialty gaskets have been developed for sealing joints where caulking may not be appropriate.

Sill plate gasket
- Polyethylene foam strips.
- Installed between the foundation and sill plate during construction or where existing house walls meet a new addition.
- Available in 152 and 203 mm (6 and 8 in.) widths on 24 m (79 ft.) rolls.

Electrical outlet and lighting fixture gaskets
- Foam gaskets designed to fit behind the cover plates of electrical receptacles, switches and lighting mounts, reducing air leakage into walls and attics.

- Electrical outlet gaskets are more effective when caulked and should be used in conjunction with child safety plugs to reduce air leakage through the electrical sockets.

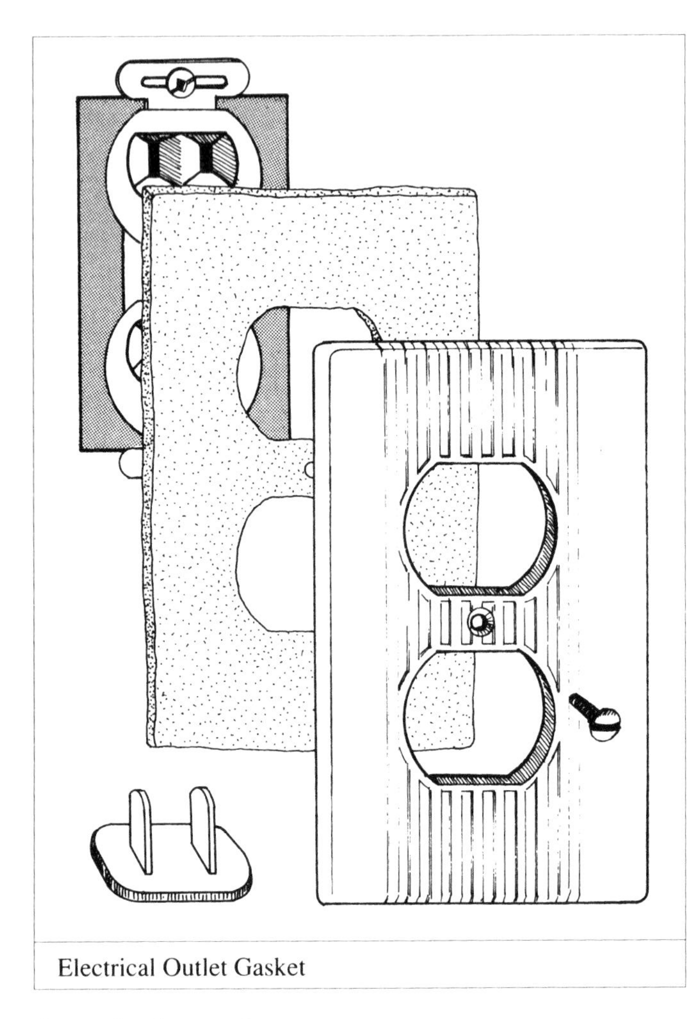

Electrical Outlet Gasket

Foam backer rod
- Closed-cell compressible foam "rope".
- Excellent for filling deep gaps before caulking.
- Available in 6.4 to 51 mm (1/4 to 2 in.) diameters.

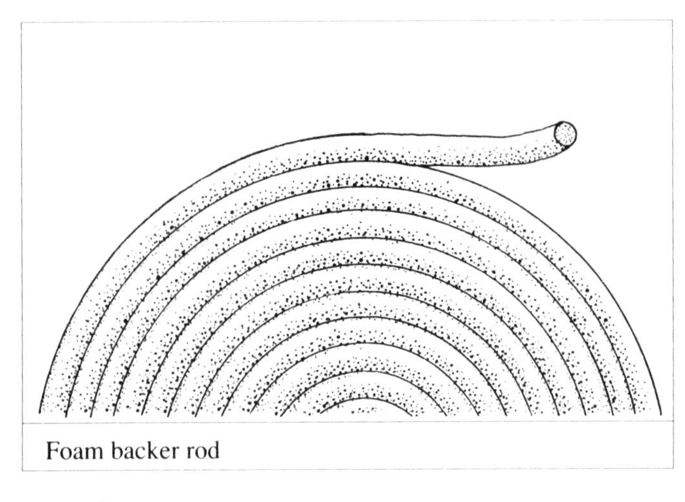

Foam backer rod

Neoprene gasket
- Flexible and very durable.
- Excellent for sealing joints and penetrations where movement is to be expected, such as plumbing stacks.

Weatherstripping

Weatherstripping is used to block air leakage aroun doors and the operable parts of windows. Weatherstripping comes in a variety of shapes; it can be a flat strip, tube or "V", and can be designed to work under compression or by sliding along the joint. **To be effective, the product must close the gap and not allow air to pass.**

Consider the size of the gap to be sealed, durability, ease of installation and appearance when choosing weatherstripping. Look for products that are flexible, and that spring-back to their original shape quickly and easily. Avoid products which make it hard to operate the window or door.

Compression Strips

Use compression strips where there is a pressure stress, such as at the bottom of vertical sliding windows, along attic hatches or on hinged windows anc doors.

Closed cell foam
- An adhesive-backed foam stripping available in rolls.
- Easy to install.
- Available as a high performance compressible polyurethane strip with its own carrier.

Ribbed closed cell rubber
- An adhesive-backed stripping available in rolls.
- Very durable, easy to install.
- Good for irregular surfaces, but has difficulty accommodating long or varied gap widths.

Tubular stripping
- Tubular material, made with either its own attachment area, or on an attachment strip of a different material.
- Rubber-type (as opposed to plastic) should be used for its better durability.
- Generally used as a window or door weatherstrip installed with nails, staples or screws depending on the type of attachment strip.
- Highly noticeable when installed.

Tension Strips

Spring vinyl

- Spring vinyl tension strips can be used in the same applications as compression strips, as well as in sliding joints in double hung windows and doors.
- Adhesive backed; easy to install.
- Two types of spring vinyl are available:
  Small format "v" strip for narrow gaps, such as tight-fitting double-hung windows.
  Large format for wide gaps, such as loose windows and doors.
- Good durability; polypropylene type should be chosen over other plastics.

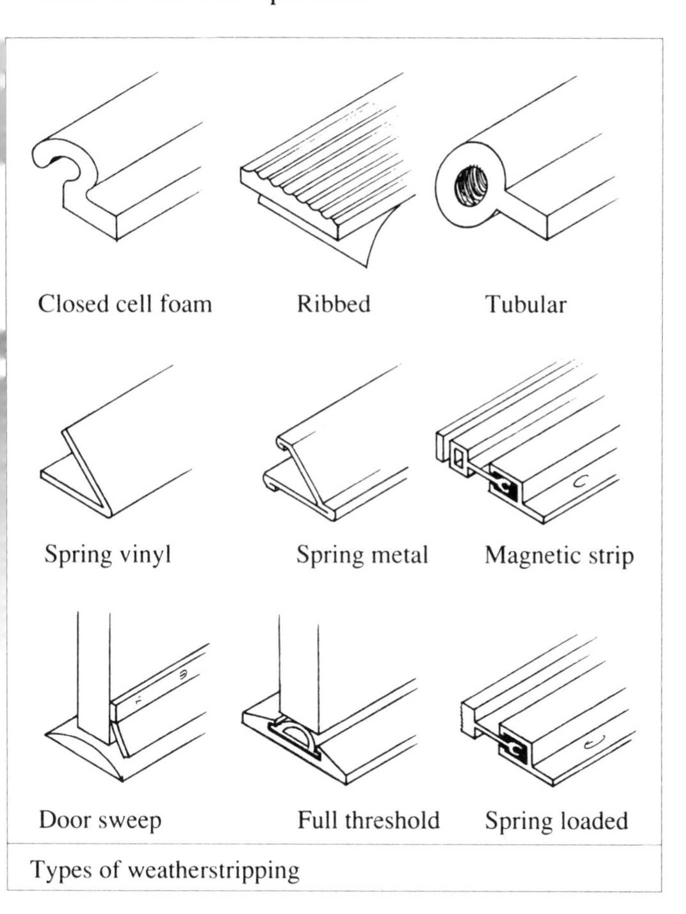

Types of weatherstripping

Spring metal

- Generally used as a door weatherstrip; most effective when under light compression.
- Installed using small tacks.
- Metal can permanently deform.

Combination Types

Spring-loaded/self-adjusting weatherstripping

- Uses a spring mechanism that allows it to adapt to unequal distances from the weatherstrip to the door or window.
- Effective for doors and hinged windows.
- Installation is by screws through an attachment strip.
- Can be used in conjunction with "V" strips.

Magnetic strip systems

- Magnetic attraction between a magnetic strip mounted on the door/window frame and a metal strip mounted on the door/window provides the seal.
- Effective for doors and hinged windows, in moderate climatic conditions.
- May not provide good seal in cold temperatures due to condensation and frost formation – PVC case may stiffen and split.
- Good durability, highly noticeable.

Door Bottoms, Sweeps, and Thresholds

The bottom of doors can be sealed using a number of systems: door sweeps, door thresholds and door bottoms. Door bottoms and sweeps are usually more durable than thresholds although they often provide a less effective seal.

Door sweeps

- A vinyl pile or rubber sweep.
- Screwed via attachment strip to door bottom.
- Effective where the carpet is low or not present.

Partial threshold

- A vinyl or rubber strip, attached to the door threshold.
- Provides an excellent seal.
- Can become damaged by traffic and weathering.

Full threshold

- A combination strip, attached to threshold.
- Requires at least 15 mm (5/8 in.) clearance below the door to be effective.

Door bottoms

- Combination strips of vinyl pile or compressible rubber.
- Attachment strip fits over door bottom.
- Requires a clearance of 8 mm to 13 mm (1/3 in. to 1/2 in.) under door.

Tape

Duct tape

- Vinyl and foil tapes can be used to seal around the seams of heating ductwork to reduce air leakage, especially where ducting passes through unheated areas of the house.

Sheathing tape

- Sheathing tape is available for sealing the seams of spunbonded olefin wind barrier material and polyethylene air barrier material.

Tape

Other Items

Electrical box air barriers

- These plastic boxes are placed around electrical outlet and switch boxes before installation.
- They are equipped with a flange for sealing to the main air barrier.
- These boxes also act as vapor barriers.

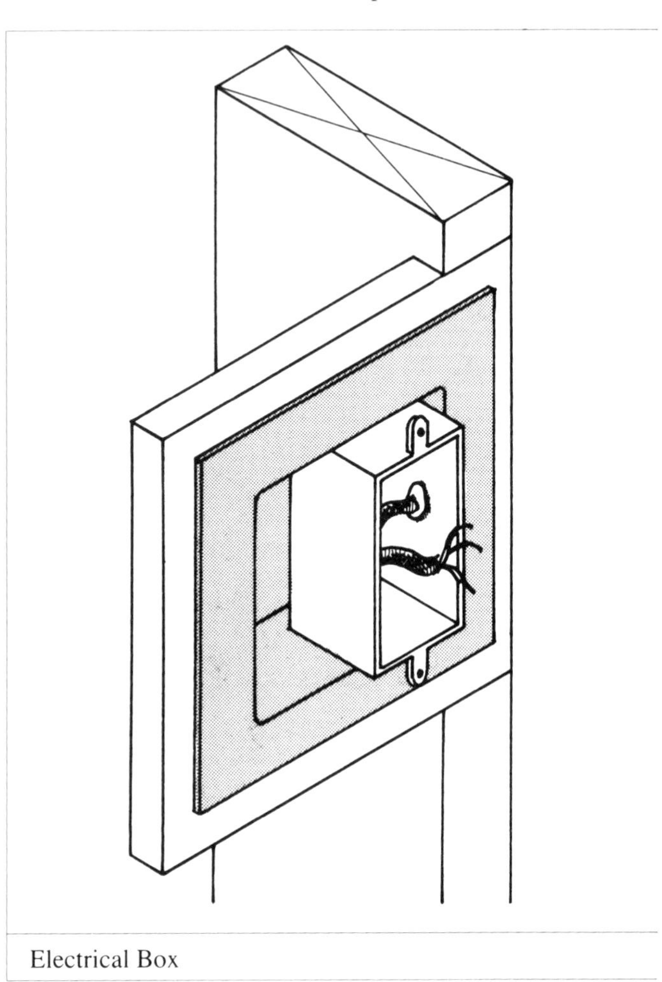

Electrical Box

Part III Vapor Barrier Materials

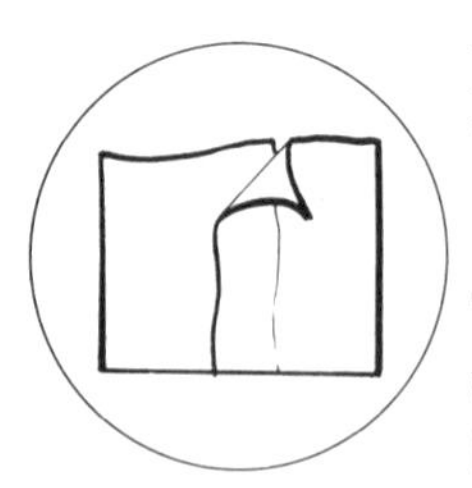

Requirements for Vapor Barriers

The vapor barrier is an important component of the house envelope; it provides some protection from moisture damage to the structure and the insulation materials. To be effective the vapor barrier must be:

- resistant to the flow of water vapor;
- durable; and
- located on the warm side of insulation.

The vapor barrier does not need to be perfectly continuous like an air barrier, but it should cover as much of the building envelope as possible. While it needs to be located on the warm side of the insulation, the vapor barrier can be part way into the wall, provided that no more than 1/3 of the insulating value of the wall is on the warm side of the vapor barrier. This should be reduced to 1/4 or less of the insulating value in very cold climates or buildings with high moisture sources such as swimming pools.

Like an air barrier, the vapor barrier can be made up of different materials; even some existing building components, such as plywood, paint, or vinyl wallpaper may form part of the vapor barrier.

Vapor Barrier Components

The effectiveness of a vapor barrier material is measured in terms of its perm rating. The lower the perm rating, the more effective the vapor barrier.

Materials that are considered to be effective vapor barriers include:
- polyethylene
- aluminum foil
- some types of paints
- some types and thicknesses of insulation
- vinyl wallpaper and
- exterior grade plywood

In most older houses the layers of oil-based primer paint and varnish finishes can function as an adequate vapor barrier for walls and ceilings. Areas that most often require special applications of vapor barriers include interlocking ceiling tiles and new drywall. Pay special attention to areas of high humidity, such as kitchens and bathrooms.

Choosing Vapor Barrier Materials

Any material used as a vapor barrier will need to be durable and resistant to moisture flow. Once these characteristics are met, other factors should be considered.
- Is it easy to install?
- Can the material also act as insulation or an air barrier?
- Is it appropriate to the other work being done on the home?

Part IV Health and Safety Considerations

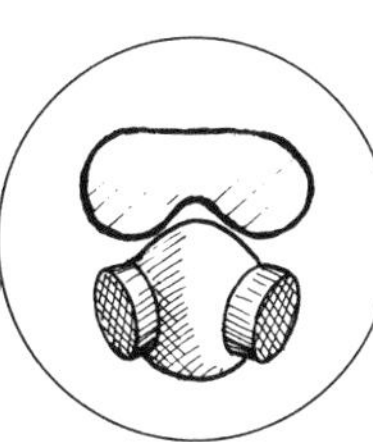

Retrofitting should pose no threat to the health and safety of the occupants or those doing the work, if proper precautions are taken. Almost all building materials are potentially hazardous, but, if handled and installed with care, the work can be done easily and safely.

Safety reminders for each type of retrofit job are noted in the 'how-to' chapters. This section provides general construction safety tips and guidelines for working with different types of retrofit materials.

General Construction Safety

- Common tools, such as hammers, utility knives, staple guns, ladders, rakes, and power tools must all be handled with care. Complex equipment, such as blowers, foamers or sprayers require special instruction and practice.
- Have a first aid kit and a fire extinguisher handy and know how to use them.
- Protect your back when lifting heavy objects; don't lift and reach at the same time.
- Take special care when handling heavy or bulky objects, especially up and down stairs and ladders.
- Smoking is especially hazardous. Don't take smoke breaks near insulation or fumes.
- Keep your work site well organized, with tools out of the way of traffic and give yourself plenty of clear space to manoeuvre.
- Make sure the workspace is well lighted and ventilated.
- Ensure proper electrical supply for power tools.
- Wear appropriate protective clothing for the job at hand.

Protective clothing

- Don't even think about working in an attic on a hot day. Heat stress can cause accidents and serious illness.

Warning: An older home may contain insulation that is wholly or partly asbestos, usually white or grayish white in colour and may be in powder or semi-fibrous form. Asbestos inhalation is associated with a wide range of cancers, even from a single heavy exposure. Before working in such material, check with your local or provincial health authority.

Protecting Yourself and Your Family

Many of today's materials give off particles, fibres or fumes that can be harmful to the installer and anyone in the immediate area. Even natural materials such as sawdust and plaster dust can be harmful. Often, the hazard is not from the primary material but from binders, solvents, stabilizers or other additives of which you may not be aware.

Fortunately, there are a number of things you can do to ensure that your retrofit job is done safely as well as effectively. Maintaining a clean work area and separating it from the rest of the house will minimize exposure to materials.

- Keep fibrous materials and materials that generate fumes well sealed until they're needed, and close the container at the end of the work day.
- Vacuum the work area daily to remove fibres and dust.
- Provide ventilation for the work area and isolate it from the rest of the house by closing doors or hanging up curtains of plastic.
- Provide extra ventilation for the rest of the house while the work is in progress and during any "curing" or drying period.

Recommended safety procedures for working with various materials are described below. Read this section carefully and follow the recommendations.

Insulation and Other Particulate Materials

Fibrous insulation materials such as glass fibre and mineral wool can easily irritate the skin, eyes and respiratory system. Long sleeves, tight cuffs, and loose, thick clothing will help minimize any skin irritations. Special barrier creams that protect the skin when working with fibrous materials are available from safety supply houses.

Wear goggles whenever there is any possibility of insulation dust coming in contact with the eyes. Eyes can easily become irritated or inflamed by brittle glass or mineral fibres and **permanent damage** can result.

Wear a mask for non-toxic particles if there is a possibility of breathing airborne particles of insulation material. Glass fibre, rock wool, cellulose and vermiculite all require a face mask and safety glasses for normal handling. The tiny fibres from glass and mineral insulations can cause respiratory tract irritation and lung inflammation. **AVOID BREATHING INSULATION DUSTS!** Well-designed, snug-fitting face masks are available through safety supply houses. Buy a supply of filters rated for the materials you are using and change the filters frequently.

Wear a hard hat to prevent head injuries and to protect your eyes and hair from insulation particles.

A vacuum cleaner is the preferred method of cleaning up fibres or dust. It is a good idea to attach an extension hose to the exhaust port of the vacuum cleaner to ensure that any particles travelling through the filter are not allowed to be recirculated in the household air. If you can only sweep up the material, wet it first to prevent particles from becoming airborne.

Vacuum clothing to avoid spreading the material around the house. Wash work clothes separately from other clothing.

Plastic Insulations

Rigid polystyrene insulation is essentially an inert material but it can shed particles when cut. Use of a face mask is recommended when cutting board stock.

Polyurethane and polyisocyanurate insulations give off harmful vapors when the rigid boards are being manufactured and when the material is being sprayed-in-place on the job site. The vapor causes skin and eye irritation and breathing difficulties, even at low levels of exposure. Residual amounts of vapor may be present with the rigid board material. Make sure the work area is well ventilated. These types of rigid boards will also shed particles when cut; use a mask as for polystyrene.

When applying the spray-in-place material, contractors take special safety precautions and use respirators. If you plan to have foam insulation installed inside your home provide additional ventilation until the material has cured.

Caulking

There is a variety of caulking materials with widely different chemical compositions. However, all caulking materials share some common characteristics.

- They all use solvents to keep the material pliable until it is installed.
- Once applied, the solvents will evaporate and fumes will be given off as the material sets or cures

Fumes from caulking can cause respiratory irritation or other allergic reactions. Make sure the work area is well ventilated and provide additional ventilation to the home during the curing period. The curing time can vary from days to weeks.

> **Note:** Follow directions when specified for exterior use only.

Warning: Older homes, especially those built prior to 1950, were often painted with lead-based paint. Caution should be exercised, especially when working with windows, doors, trimwork, wood siding or porches. For further information, you can obtain a copy of the publication entitled: "Lead in Your Home" (publication number: NB6624). Contact Canada Mortgage and Housing Corporation, 700 Montreal Road, Ottawa, Ontario K1A 0P7 or their nearest regional office.

Part V Special Health Considerations

Retrofitting poses special potential health problems for those with allergies, asthma or chemical sensitivity. However, there are several options that make retrofit possible for people in this situation.

- Choose materials carefully.
- Take extra precautions when working with the material.
- Plan the work to minimize exposure.

Some materials are less troublesome than others. By careful choice of materials, exposure to irritating substances can be avoided or reduced. For example, rigid board insulations don't shed dust or particles unless cut and some caulkings have a shorter curing period. Also, finishing materials such as paints and stains with reduced toxicity are now available for the chemically sensitive.

Extra care when working with the materials can also reduce exposure for you and your family. Segregate the work area, using sheets of plastic if necessary and don't wear work clothing in other areas of the house. Keep the work area clean, vacuuming frequently. Store materials outside of the house until they are needed and keep caulking tubes, insulation bundles and paint cans closely covered when not in use. Provide additional ventilation to the work area and to the whole house while work is in progress and during any curing period.

Health considerations may be a major factor in the decision to insulate from the outside of the house instead of the inside. If you do work from the inside, plan the job so that it's over and done as quickly as possible. This may mean hiring a contractor to do all or part of the work or staging an old fashioned work bee. Where major renovations and retrofit will affect the whole house you may wish to consider sending the family on a vacation or moving to temporary quarters while the work is underway.

If you have special health concerns you may also wish to seek advice from your allergist or a family doctor. They may be able to help you select materials that are more easily tolerated. For the chemically sensitive, this may involve a series of exposure tests to small samples of the material. Your allergist may also be able to direct you to contractors who are experienced in doing work for clients who are allergic or chemically sensitive.

3 Comprehensive Air Leakage Control

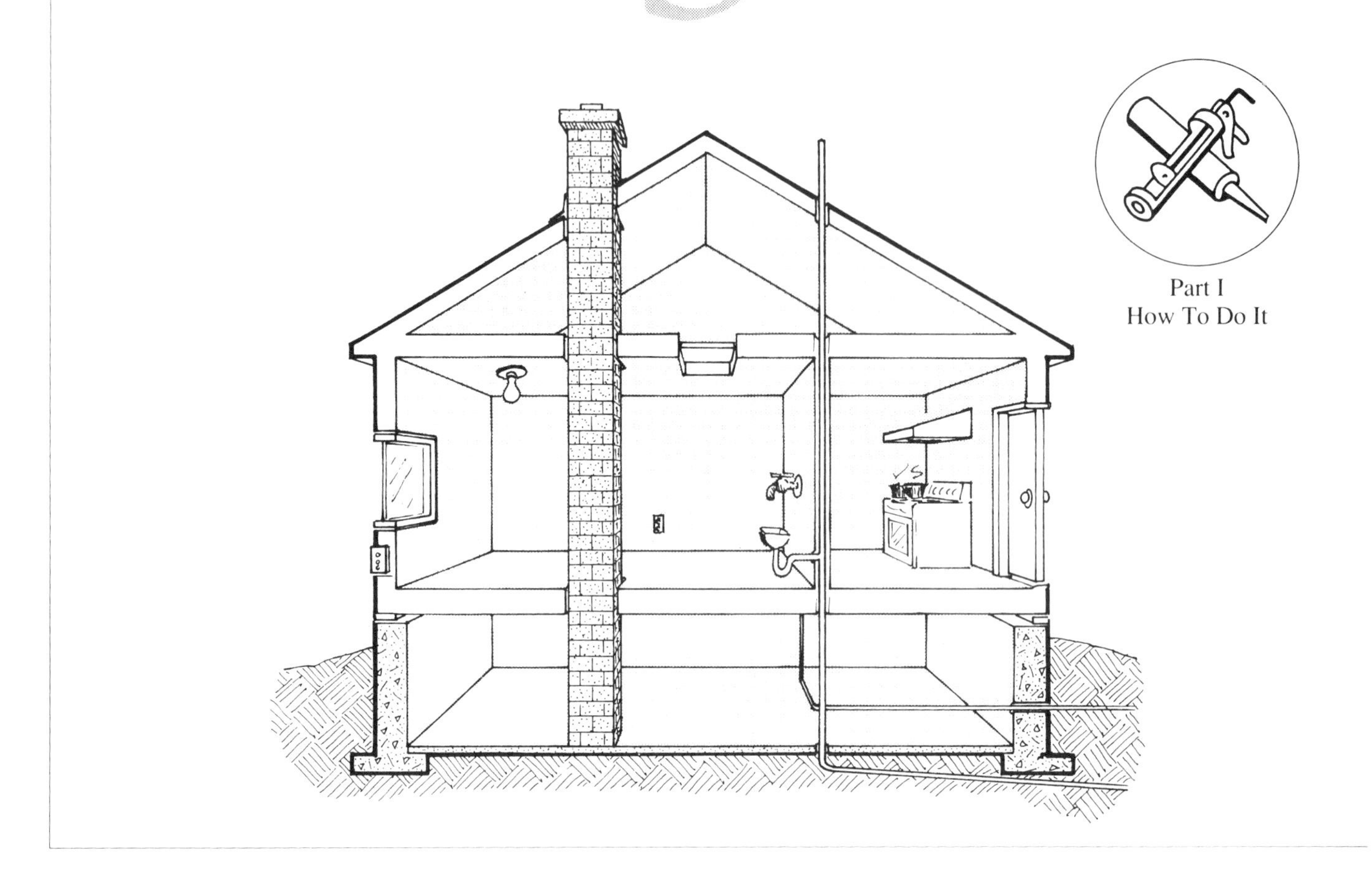

Introduction

Air leakage control is the single most important retrofit activity and it should be considered first in any retrofit strategy. Blocking air leaks brings many benefits, including increased comfort, reduced heat loss, protection of the building structure and reduction of the amount of noise and dust that enters from outdoors.

Comprehensive air leakage control is a systematic identification and sealing of as many of the air leakage paths as possible, and includes a system of ventilation. Many of the leaks are obvious breaks in the air barrier system—such as through and around windows, doors, and electrical outlets. Other air leaks are more difficult to identify, such as bypasses around chimneys and plumbing stacks that can channel air directly from the basement to the attic.

Comprehensive air leakage control can result from a systematic effort of weatherstripping, caulking, and the application of gaskets and tapes. Air leakage control is also an essential part of every insulating job. Every time you insulate, you must also install or upgrade the air barrier system. This will help you to get the most from your insulation job and help ensure that moisture does not settle inside the insulation or building envelope.

Warning: Furnaces, fireplaces, wood stoves, and any other fuel-burning appliances also require air for combustion and for diluting and exhausting the products of combustion out of the house. If there is not enough air, it is possible that the chimney or flue could backdraft or spill dangerous gases into the house. Refer to the section on "Combustion Air" on page 97 for further information. Also, if you suspect that you have a problem, you should speak to your heating contractor.

House As a System

Remember that the house works as a system, where each component is related to other components. Changing one thing can affect other aspects of the house.

This is especially true with air sealing which can affect the house moisture flows and combustion and ventilation air supply. As the envelope is tightened, the household humidity levels rise. This can cause condensation and moisture problems. Less air is available for combustion appliances and there is less fresh air. Therefore, an important part of comprehensive air leakage control is attention to whole house ventilation and combustion air supply.

Each house will respond to comprehensive air sealing in its own unique way. This has to be monitored in each case. For example, a large old house may be very difficult to seal to the point where problems occur, while other houses may require remedial measures even before comprehensive air sealing. The best way to avoid problems is to understand how they occur and to take steps to control humidity and ventilation.

These issues of humidity, ventilation and combustion air are discussed in more detail in Chapter 8.

Part I How To Do It

How to Locate Air Leaks

The first step is to identify where the leaks occur. Air leaks anywhere there is a hole in the building envelope and a pressure difference. During winter, the house tends to operate like a big, fat chimney. This means that air tends to enter the house at lower levels, and exit at the upper levels and ceiling.

Identifying the specific leakage areas can sometimes require a little detective work. You can hire a contractor who has the right equipment to do a test for you. Or you can do your own detective work by making a "draft detector", and using the checklist of leakage areas and perhaps your own "pressure test" to locate those places where air is leaking.

Leak detector

There is an easy way to locate air leaks: make yourself a leak detector. All you need are incense sticks. Hold two or three together for more smoke and easier detection. Powerful leaks will cause the smoke to dissipate and the tips of the incense to glow. Slower leaks will cause the smoke to trail away or move towards the leak.

On a cold day, check for drafts in all suspected areas. It is easier to locate air leaks on windy days. You'll be surprised to discover how many spots need to be sealed with caulking or weatherstripping. You should also check for possible leaks on the interior walls and features of your house. There may be a direct route through partition walls or along floor joists to the outside that should be sealed.

Leak detector

Pressure test

Professional air sealing companies often use a depressurizing fan test to identify and measure the air leaks in a house. A powerful fan is inserted in a doorway and all intentional openings—windows and doors, chimneys, vents, etc.—are closed or sealed. The fan depressurizes the house and leaks are easily identified where air rushes into the house. A professional fan test can also determine the total leakage area in the house, the extent of the work required, the effectiveness of the work as indicated in a "post-retrofit" test, and indications of back-drafting and spillage problems.

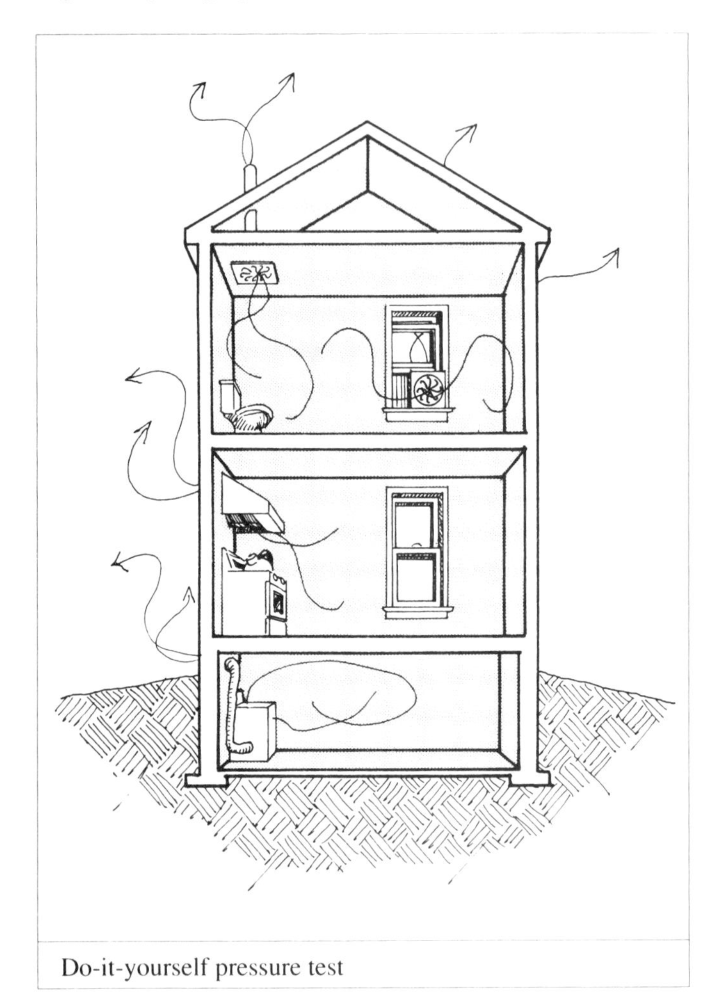

Do-it-yourself pressure test

You can perform your own rudimentary fan test by closing up all windows and doors and turning on all the exhaust appliances in the house. This would mean turning on bathroom and kitchen fans, clothes dryers (on cool cycle), and any portable fan placed in a window (if you can seal around it). Be careful during the test. Turn off the furnace and water heater (if fuel-fired) to prevent backdrafting caused by the other exhausting devices.

You can now go around the house with your leak detector and identify and mark the air leakage locations that should be sealed.

The rest of this chapter looks at caulking and installing air and vapor barriers. See Chapter 7 for a discussion on weatherstripping windows and doors.

Checklist of leakage areas

A few areas of the house deserve special attention, but don't limit your detective work to just these places.

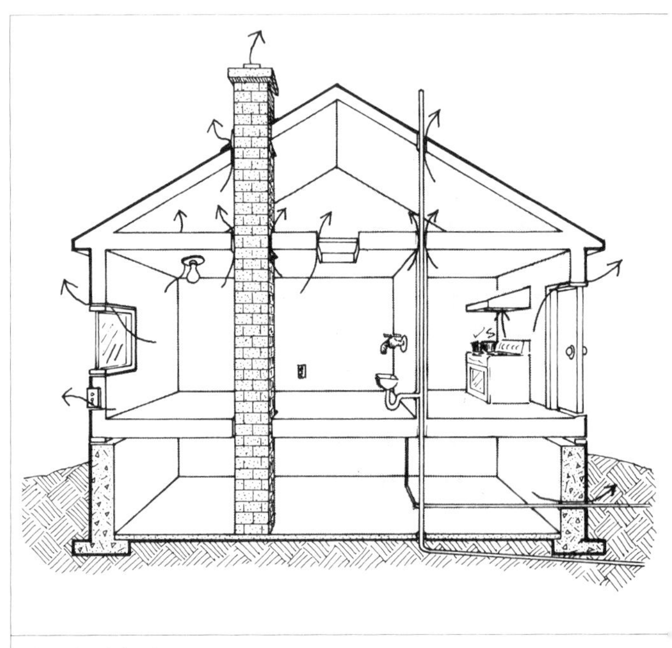

Typical leakage areas

From inside the main living areas check:

- Windows. Check the glass panes for tightness, and around both the window sash and the window casing.
- Doors. Check around the door, including the threshold and around the door frame.
- Electrical outlets, including ones on interior walls.
- Exhaust fans and vents. These should vent to the outside and close properly when not in use.
- Corners where two walls meet with an imperfect seal.
- Light fixtures in the ceiling.
- Interior trim and baseboards.
- Cracks in the wall finish or ceiling.
- The joint where a wood frame wall joins a masonry wall or chimney.
- Doors and hatches into unheated attics.

- Fireplace dampers and fireplace bricks.
- Behind bathtubs and under sinks.
- Above sliding pocket doors.
- Around plumbing pipes and ductwork.

From inside the attic check:
(you may have to move aside existing insulation)

- Around the plumbing stack and any other pipes entering the attic.
- Around wires or ceiling light fixtures that penetrate the attic floor.
- Around ducting that enters the attic from inside the house.
- At the junction of the ceiling with interior wall partitions.
- Attic access doors.
- Around chimneys.
- Along any shared walls.
- Ceiling area over bathrooms and stairwells.

From inside the basement check:

- Where the wood frame wall (sill plate) meets the masonry (concrete or stone) foundation or where joists penetrate the masonry wall.
- Any holes or gaps where the electrical lines, gas lines or oil fill pipes go through the wall (careful!).
- Holes for wiring and plumbing going into external walls.
- Leaky ducting, or poorly fitted hot air registers or cold air intakes.
- Around window and door framing.
- Cracks in the foundation wall and slab.
- Floor drains.

Caulking

Air seal any cracks and penetrations on the **inside** surface of exterior walls, ceilings or floors. Interior sealing will prevent air from escaping into hidden cavities in the walls and roof. The sealing will be protected from the elements on the inside and it will be easier to periodically check its condition. Any moisture that does reach that wall space, however, should be allowed to escape to the outside or moisture problems might result.

It is not advisable to seal the outside surface of an exterior wall (i.e. the cold side). Caulk only those cracks that will allow water entry. If you are painting the house, try not to plug the joints in the siding and use a permeable (latex) paint or stain. The outside of the walls must be left alone to breathe and dispel any moisture.

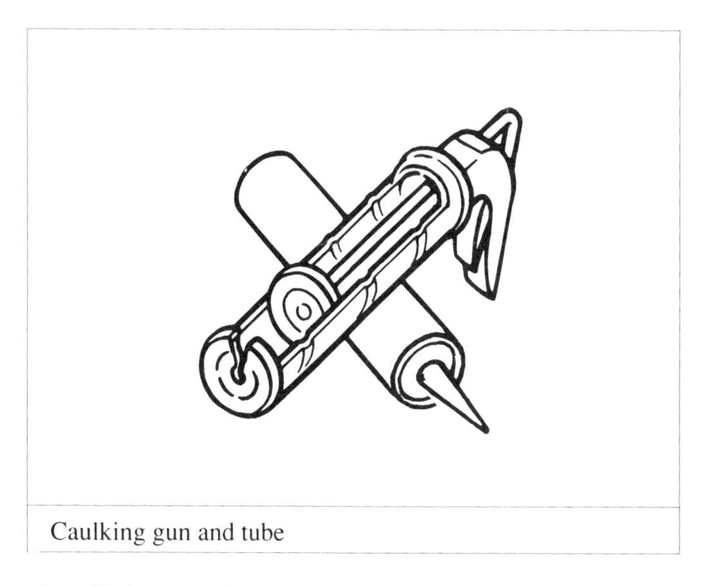

Caulking gun and tube

Caulking — What It Is and How to Apply It

Use tubes of caulking compound and a **caulking gun** for most sealing jobs. Try the grip with a tube in the gun before buying. Some guns may have a trigger spread too large for your hand. The gun should have a shut-off. A thumb release on the gun is a big convenience since it permits one hand operation.

If you are doing a complete job, you will need many tubes of caulking — one standard size tube will make a bead 6 mm (1/4 in.) wide and 7.6 m (25 ft.) long.

There are a number of types of caulking compounds available. People often have a bad experience when they first try caulking because they purchase an inexpensive or inappropriate caulking compound. The very cheap compounds are not only more difficult to apply, but lack durability. Be sure to choose a material well suited to the task! See the previous chapter for a guide to the different types of caulking and their application.

Note: If you plan to caulk around items which are a source of heat (chimneys, light fixtures, fan motors, etc.) be sure to use a **heat-resistant** caulking compound. Silicone or polysulfide sealants usually work well. Special high temperature silicones are available for flue pipes.

How to Caulk — Step-by-Step

1. Identify the areas to be caulked (as outlined earlier in this chapter).

2. Do not try to caulk in an area where the temperature is below about 5°C as the compound will become stiff and difficult to work with. **Read labels carefully for storage and application temperatures.**

3. Make sure the area to be caulked is clean and free of dirt, loose paint, and old caulking. Replace deteriorated wood and renail loose boards. If there are particularly large cracks (greater than 6 mm (1/4 in.) use a special filler such as oakum or a foam backer rod before caulking. Push this material into the crack to a depth equal to half of the crack width.

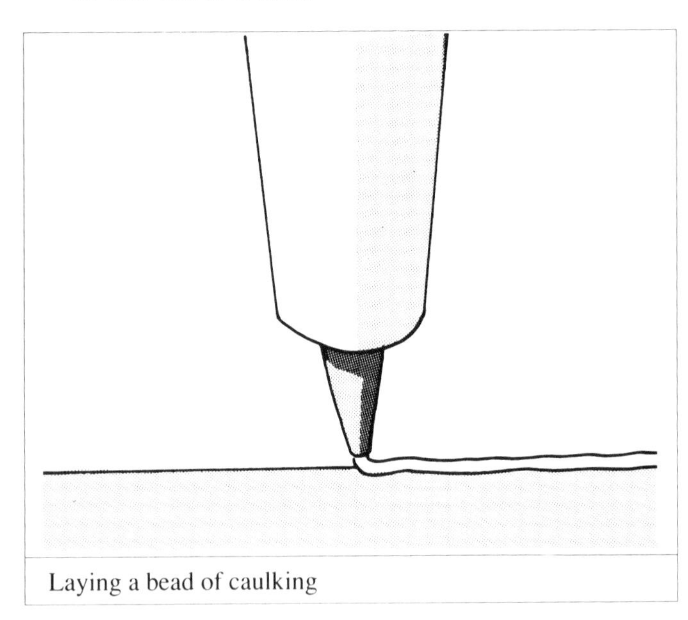
Laying a bead of caulking

4. Cut the nozzle of the tube at a size that will allow the bead of caulk to overlap both sides of the crack. Make the cut square and then break the seal with a wire or long nail pushed down the nozzle.

5. Push the caulking gun along at right angles to the crack or joint. The caulk is then forced into the crack to fill the gap completely. Make sure the caulk adheres to both sides of the crack and that there is sufficient caulk to allow for movement or shrinkage.

6. It should also look good. Effective caulking takes practice, so go slowly at first, following instructions on the tube. "Tooling" or finishing the bead can usually be done with a wetted sponge or finger before the caulk sets, but don't use your mouth to wet your finger!

7. Latex and silicone caulk can be cleaned off with water before they set. For other caulks you can try a standard solvent (e.g. toluene, varsol, or brush cleaning solvent) or check the manufacturer's literature. Be sure to release the pressure lever on the gun to prevent dripping as you move from place to place.

Other Sealing Materials and Applications

A number of other materials are used to provide an air barrier at different locations in the house. These include specialty gaskets and tapes, as well as sheet materials, such as polyethylene, spun-bonded olefin rigid insulation, drywall, plywood, and sheet metal. Installation techniques are critical when using sheet materials as an air barrier. All edges, seams and penetrations in the sheets must be sealed. Further details are provided in other chapters.

Air and Vapor Barriers

During renovation it is often possible to install a new air and vapor barrier using sealed drywall as the air barrier and layers of paint or sheet polyethylene as the vapor barrier. Alternatively, sealed sheet polyethylene can be installed on the warm side of the insulation to provide both an air barrier and a vapor barrier. Proper installation is critical.

- Use wide sheets to minimize seams.
- All seams and edges should overlap over a solid backing, such as a stud.
- Run a bead of non-hardening acoustical sealant between the overlapped sheets over the support.
- Staple through the sheets and the bead of sealant. All other staples should be avoided or minimized.
- The finish (e.g. drywall) acts as an anchor, securing the seam. If the polyethylene is recessed in the wall, a batten can be nailed over the seam to provide the mechanical support.
- Seal all penetrations. Where possible, they should penetrate at a solid backing such as plywood and be caulked.

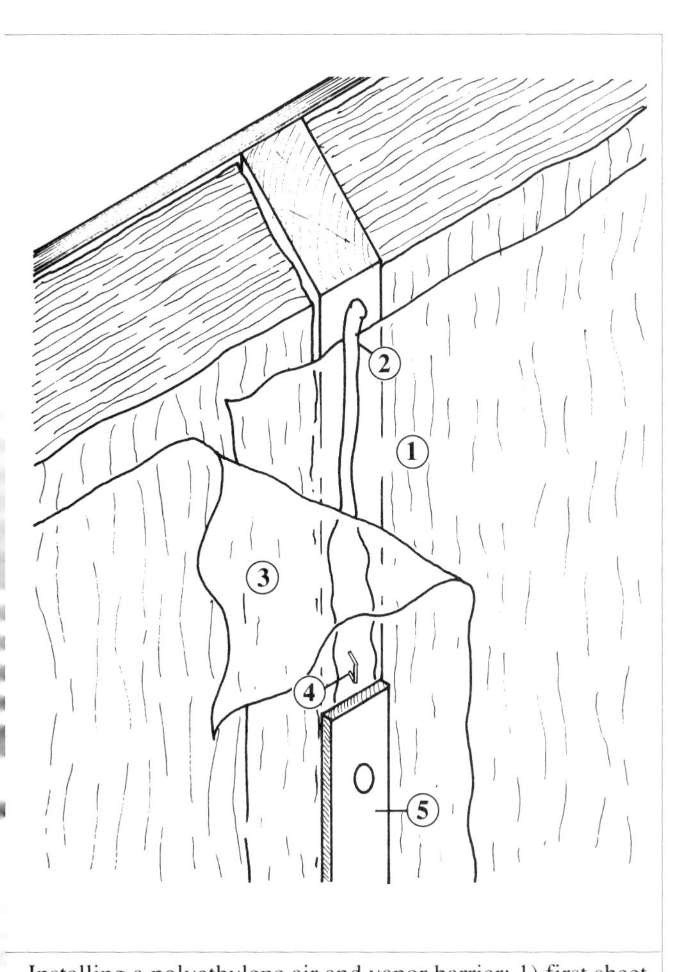

Installing a polyethylene air and vapor barrier: 1) first sheet over a solid member; 2) bead of acoustical caulking; 3) second sheet pressed into bead; 4) staples through bead; 5) wallboard or batten for mechanical support.

Tips on How Best to Seal Some of the Leakiest Areas

Electrical Outlets

If you notice a draft through an outside wall electrical outlet, it **must** be sealed. (Some inside wall outlets can also provide leakage paths—check them.) Turn off the power to the outlet by turning off the circuit breaker or removing the fuse. Check to make sure the power is disconnected by plugging in a lamp. There are special CSA approved **foam pads** that fit between the cover plate and the receptacles. You'll obtain a better seal if you caulk the gasket before installation. Place child **safety plugs** in seldom-used outlets. Some foam pads come with a gasket that fits on the safety plug.

If you are installing an electrical outlet during renovation, get a good seal by placing it in a special plastic box that is available from many hardware or electrical supply stores. Caulk the penetration for the wire and seal the new air and vapor barrier to the edge of the box.

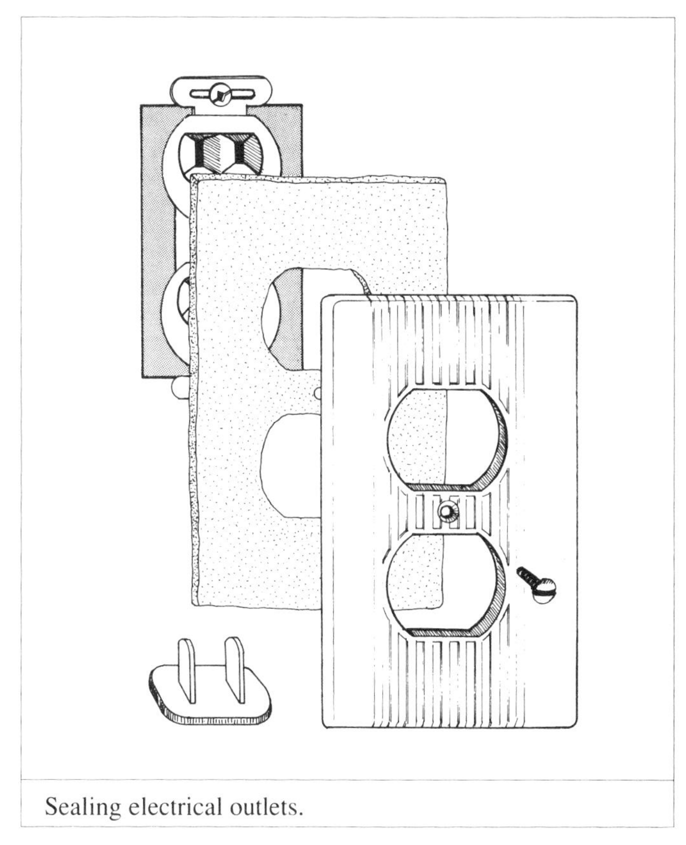
Sealing electrical outlets.

Trim Areas (Baseboards, Mouldings, Window and Door Casings)

Seal areas of air leakage around all trim. In some cases this can be done easily by sealing all the joints with a flexible caulk that is clear or paintable, or of matching color. A more effective solution for leaky or poorly fitted trim is to carefully remove the trim and seal behind it. Insulate wide cracks with a foam backer rod and seal them with caulking, polyurethane foam, or other suitable material.

If the baseboards are removed, you might also be able to caulk between the wall finish and the bottom wall plates, and between the plates and the floor.

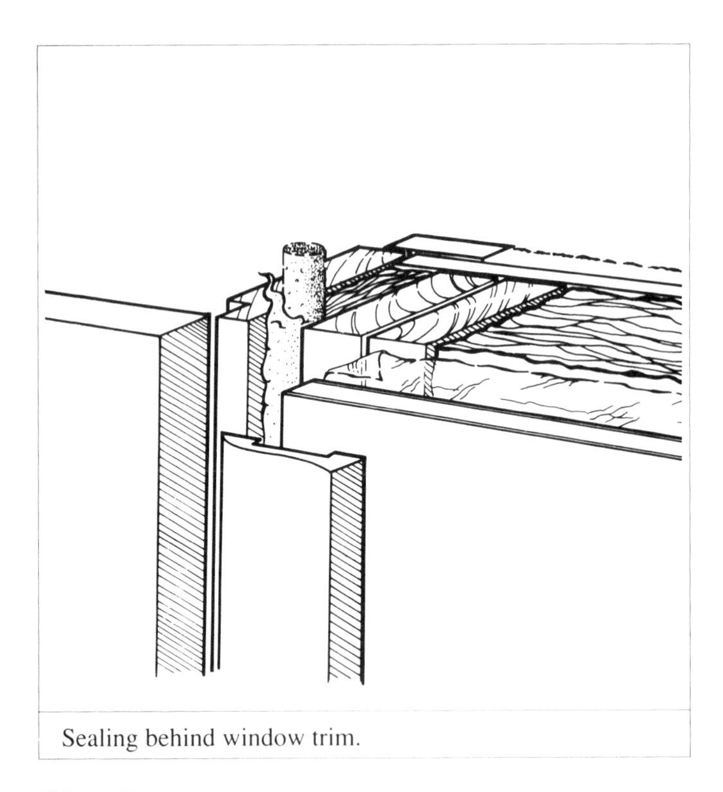

Sealing behind window trim.

Glass Panes

The seal between glass and its wood frame should be tight. Check the glazing carefully and be certain that all the seals are intact, with no cracks or missing sections. If not, repair them with *putty* or *glazing compound*. Putty usually costs less, but tends to dry out and crack faster, unless you put linseed oil on the wood first. Glazing compound, on the other hand, lasts longer and stays semi-soft and usable for longer. Remove the old putty and apply the new material with a putty knife. Be sure to press it firmly into the space for a good seal.

Fireplaces

A crackling fire makes a room cozy, but what happens when the fireplace is not in use? If the damper is left open, warm air from the room shoots up the chimney. **When the fireplace is not in use, close the damper**. Take a flashlight and make sure the damper fits tightly. If it doesn't, fix it yourself or have it repaired.

Even with the damper closed, a great deal of heat still escapes up the chimney. Try designing an insulated fireplace plug for use when there is no fire. Commercially available **glass doors** for the fireplace are usually not very tight or effective. Ideally, you should also install an outside combustion air duct to the fireplace to improve operation, efficiency, and safety. Ask at your local building suppl outlet or wood-burning appliance dealer whether a kit is available.

Seal off an unused fireplace. Close it up by putting an air-tight plug of some sort in the chimney or across the fireplace opening. This can be made fron board material that is cloth covered and provides a good seal at the edges.

Check for any air leaks where the chimney meets the wall (remove the trim if necessary). Caulk this joint with a flexible caulk.

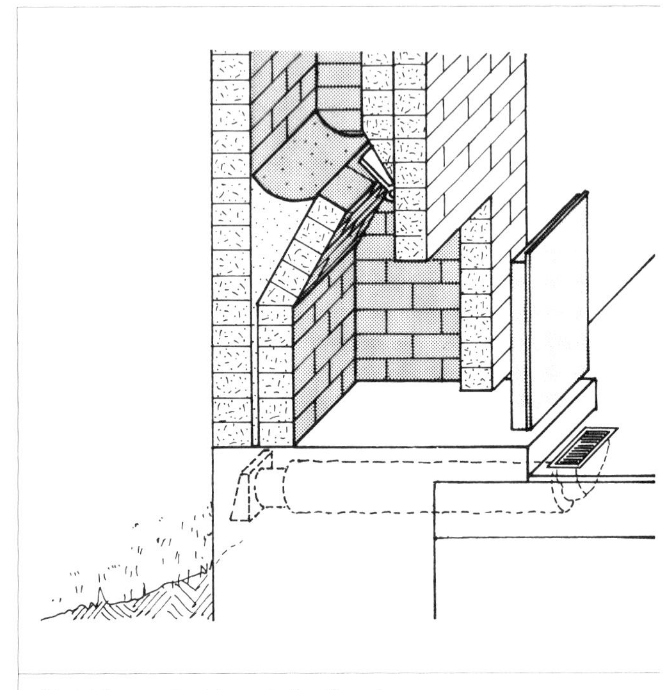

Outside combustion air for fireplace.

Chimney

There may be a large gap where a masonry chimne rises through the attic. This space can be partially sealed by stuffing it tight with pieces of mineral wool batt. **Do not use any material that is, or ma become, flammable**. For greater effectiveness, cut two pieces of sheet metal to fit around the chimney Seal all the joints with a flexible, **heat-resistant** sealant.

If you have a factory-built metal chimney rising through the attic, do not insulate closer than 50 mm (2 in.) as this can create hot spots in the chimney lining and presents a fire hazard. Instead, install a collar of metal or other fire-resistant material aroun the chimney and caulk to prevent air leakage into the attic.

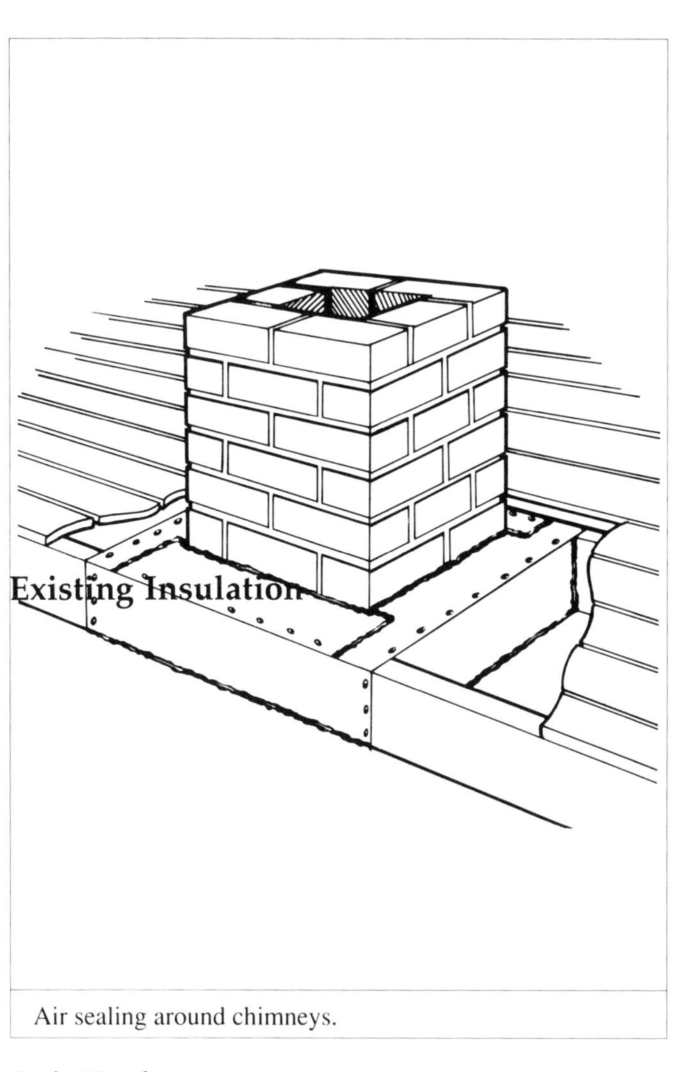

Air sealing around chimneys.

Attic Hatch

Seal the attic hatch exactly as you would seal a door to the outside. Caulk around the frame and between the casing and the ceiling plaster board. Apply weatherstripping along the edges of either the casing or the access panel itself.

Finally, install hooks with eye bolts, or some sort of latch mechanism to hold the hatch firmly against the weatherstripping. The hatch itself should be insulated.

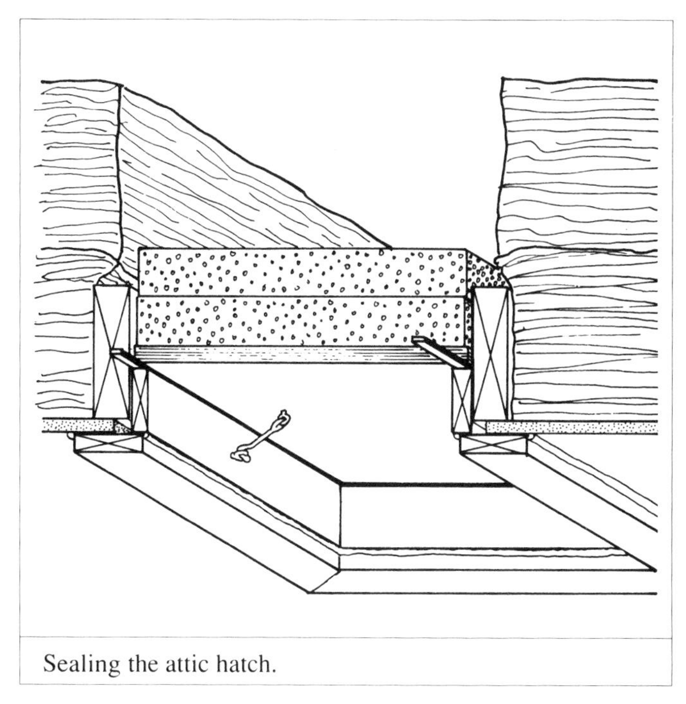

Sealing the attic hatch.

Windows That Are Never Opened

If you don't need a window for ventilation or as a possible emergency escape route, seal it by **caulking** rather than weatherstripping. Caulking is likely to be quicker, cheaper, and more effective. First, close the window sash onto a bead of caulk. Pry loose the vertical stops on each side of the window with a putty knife. Start prying at the centre, and work towards the top and bottom. (If you are careful not to bend the nails, replacement will be easy.) With the stops removed, caulk between the sash and the window casing. The caulk can then be concealed by the stops. Alternatively, install a special strippable caulk that can be removed when you want to operate the window again.

Mail and Milk Chutes

Seal the chute if it is no longer in use. If you use it regularly, seal around the frame and replace the weatherstripping. If the chute cover is on a spring, make sure it closes properly. A drop of oil can sometimes work wonders or, if necessary, replace the spring. Alternatively, you can buy a new cover. Consider closing your mail chute and replacing it with an outside mailbox.

Contracting the Work

Homeowners can usually do an effective air sealing job if they have the time and patience, and are conscientious about air sealing in areas that can be difficult and uncomfortable to work in (e.g. the attic). However, professional air sealers can usually do a much better job because of their experience in locating and sealing leaks. Contractors may be experienced in using depressurizing fan doors and smoke pencils, and specialty caulking and sealants.

Another service many air sealing companies can offer is testing and assessment of ventilation and combustion air requirements, including backdrafting testing.

Typical whole house air sealing can cost from $500 to $2000 depending on the size and complexity of the house. The contract should specify each area to be sealed and the materials to be used.

Roofs and Attics

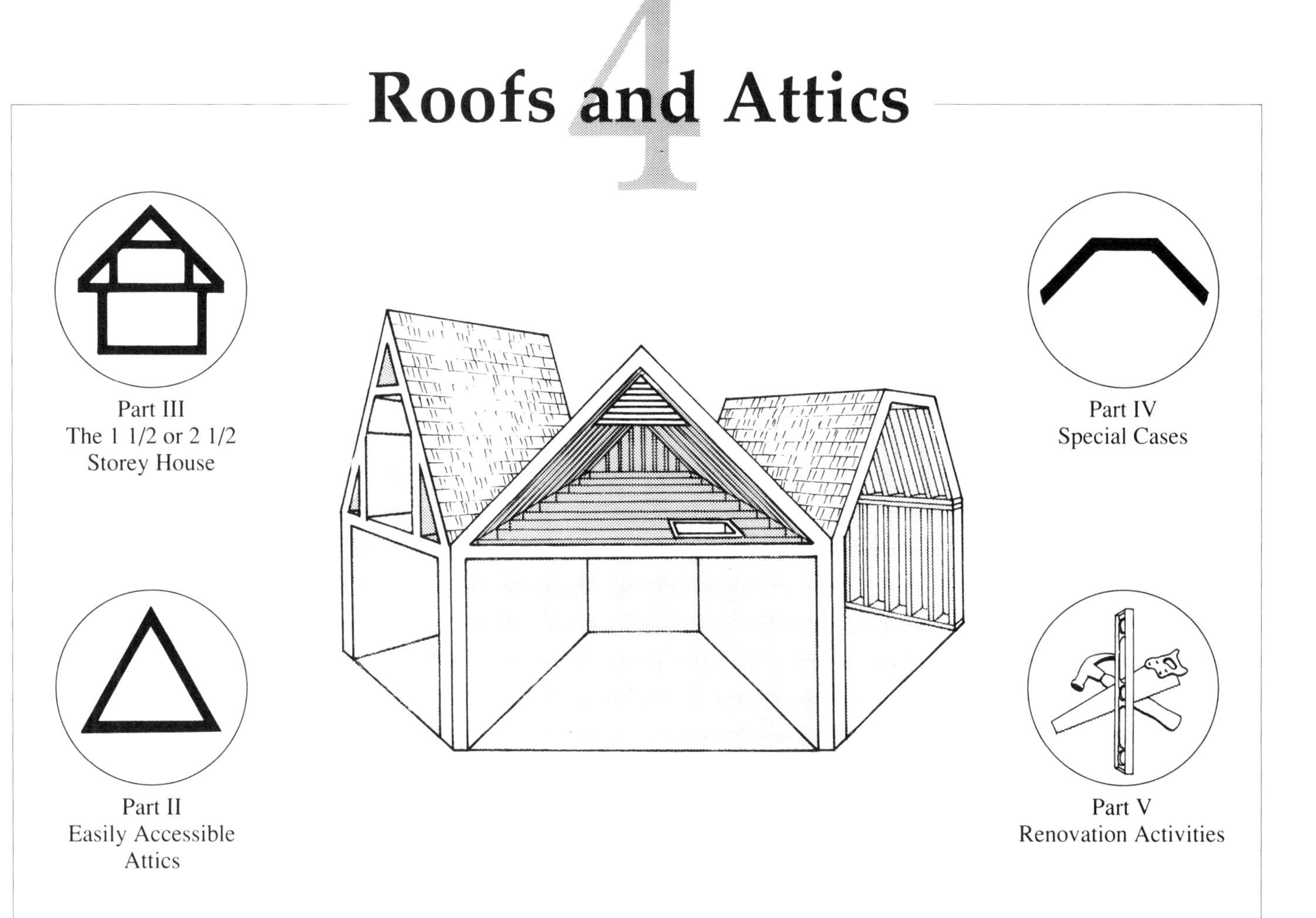

Introduction

When you think of adding insulation, the attic is the first place that generally comes to mind. Relatively easy access and few obstructions have made the attic a favourite starting point for many homeowners. This is despite the fact that most other areas — such as basements and uninsulated walls — lose more heat than the typical attic.

Adding more insulation is only half the battle. Even if an attic is already insulated, there may still be an opportunity to improve the energy efficiency and soundness of the house through air sealing. Air leaks into the attic through the many cracks and penetrations can account for substantial heat loss, and can lead to a variety of moisture related problems. **The importance of air sealing cannot be overstated.**

This chapter examines methods for insulating and air sealing a variety of attic types.

- Part I examines some general conditions that apply to most attics.
- Part II will deal with easily accessible attics.
- Part III will deal with houses that contains a half storey.
- Part IV will describe retrofit work in inaccessible attics, cathedral ceilings and flat roofs.
- Part V will deal with renovation where the attic is converted into living space, a new roof is added or where the interior is to be refinished.

Part I General Considerations for All Attics

Whatever the attic or ceiling type, there are a number of things to examine before beginning work. A thorough inspection of the following features will help you develop your retrofit strategy.

Accessibility

Most houses with accessible attics have an interior ceiling hatch, although an exterior roof or wall-mounted entry is not uncommon. The hatch should be large enough to allow you to bring in materials. If this is not the case, or if there is no access, then an entry should be installed. Exterior-mounted entries represent one less opening that will have to be air sealed. However, interior hatches are more practical if you want to inspect the attic during the winter months.

Check the roof space for obstructions and ease of movement. Vertical clearances of less than 1 metre (39 in.) won't let you move freely. Attics or roofs without a working space are covered in Section IV.

Assessments

Check the condition of roof framing, sheathing, finish, and the soffit and fascia for signs of moisture problems such as leaks, stains, mould, flaking or rot. Unchecked moisture problems will reduce the effectiveness of your insulation and can lead to structural damage such as wood rot or split rafters. Mold, mildew, fungal growth or rot are sure signs of advanced condensation problems. The cause will have to be identified and fixed before adding insulation.

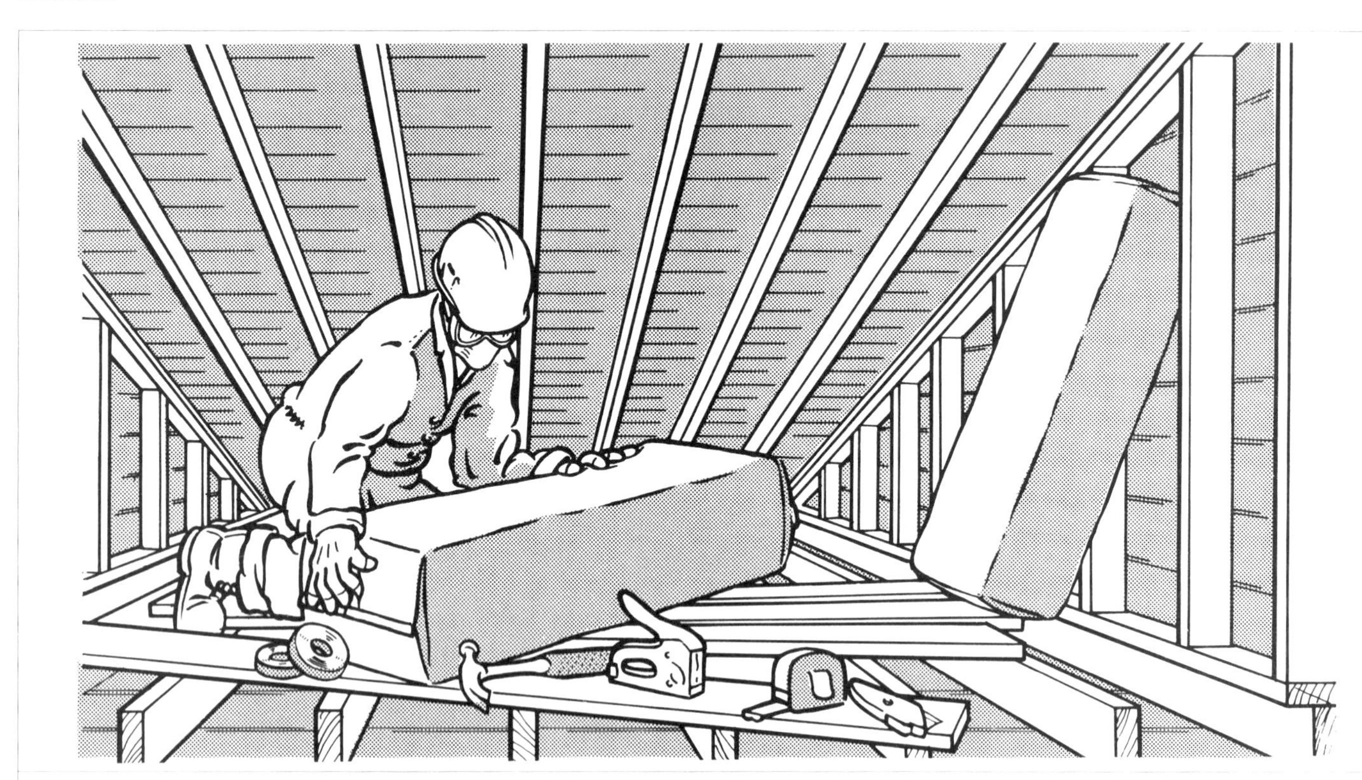

Some attics are easier to work in than others.

Moisture can come from the outside due to failure of the roof or flashing. Typical problem areas include poor flashing at a hip or valley, or at the chimney. Pay particular attention to water marks on the underside of the sheathing or along rafters.

Moisture can also come from inside the house, carried into the attic as water vapor by air leakage. Typical problem areas occur around bathroom and kitchen vents that penetrate the ceiling, around plumbing stacks and chimney chases and at wiring penetrations and pot lights.

Check the attic in late winter for condensation build-up, which will appear as frost in cold climates. Checking your attic during or just after a rain storm will help determine whether moisture problems are generated by interior or exterior sources.

Wiring

Check that any electrical runs in the attic are in good condition. Any wires with frayed or torn insulation will have to be replaced before insulating. Make sure electrical connections are free of rust and corrosion. **Work safely around electrical wiring!**

Recessed lights should be removed and replaced with surface mounted fixtures. Recessed lights are difficult to seal effectively, and can be a fire hazard. However, there are CSA-approved recessed fixtures that can be safely installed. Seek advice from your local utility about how to air seal and insulate around them.

Examine the existing insulation for:

- condition (dry, wet, compacted, etc.);
- type;
- average depth; and
- coverage.

If the insulation has been damaged or is likely to pose a health risk it should be removed. If it is wet, don't cover it until the source of moisture is removed and the insulation is dry.

If the insulation is dry it will probably be okay to leave it in place. Generally, there is little problem in using two different types of insulation.

Check the depth of the insulation to determine its insulating value. Compare this with the recommended insulation values on page 11.

Check to make sure that the insulation is distributed evenly and that there is full depth coverage. This is particularly important around the perimeter of the attic above the wall plates. Uninsulated areas will cause a cold spot where the wall and ceiling meet which can lead to moisture problems. Install depth indicators – e.g. a piece of wood nailed perpendicular to a joist – throughout the area to be insulated to ensure a consistent depth of application.

Existing Air and Vapor Barriers

Most houses have a vapor barrier on the warm side of the insulation. In newer houses a polyethylene sheet usually serves as the barrier. In older homes the vapor barrier might have been provided by wax paper, kraft paper-backed batts, or layers of paint. However, very few houses have an effective air barrier, although some houses built in the last ten years may be tighter. If there is an air barrier, your job will be made much simpler. Locate the barrier and determine its condition. Remember, an air barrier must be continuous; holes or tears will have to be repaired, and penetrations through the barrier will have to be sealed.

Increased insulation means a colder attic, which in turn means that any vapor that does escape into the attic can condense before it can be vented. It is essential to air seal the attic to prevent moisture from getting in.

If there is no air barrier, concentrate your efforts on comprehensive air sealing. You can create an effective air barrier by using caulking, gaskets and weatherstripping to seal the joints between building components.

Air Sealing

One sure sign of leakage is stained insulation or frost in the winter. The staining is the result of dirt which has been trapped by the insulation as air escapes from the house. Sealing tips for common sources of air leakage are described below. It may help to make a map of where the light fixtures and walls are before you go up.

- **Around the plumbing stack** and any other pipes entering the attic. Since the plumbing stack moves up and down due to thermal expansion, the seal must be airtight yet allow for movement. This can be accomplished by using a rubber gasket, a plywood collar in conjunction with an expansion joint, or a sleeve made of polyethylene.
- **Around wires or ceiling light fixtures that penetrate the attic floor**. Seal holes where wires penetrate partition wall top plates with a compatible sealant. Some sealants damage the plastic or rubber coating on electrical wiring. A possible choice is acoustical sealant.
- **Around ducting** that enters the attic from inside the house; for example, kitchen exhaust fans, bathroom vents, etc. Seal joints in the ductwork with duct tape. It is especially important that no exhaust fans discharge into the attic. They should discharge to the outside, but not directly below the eave vents. The ducts should stay below the insulation or should be wrapped with insulation.

- **At the junction of the ceiling and interior wall partitions**. Pull back the insulation to locate cracks formed along interior walls. Seal the cracks with a caulking compound. This area will not usually pose a problem if the interior ceiling has been plastered.
- **At the top of interior and exterior walls**. Check to see if all the wall cavities are blocked from the attic (usually by a top plate). If the spaces have been left open, or cut open, install a piece of rigid board insulation in the exposed cavity. Remember to caulk the edges of the rigid board. If the top plate is cracked or poorly fitted, use caulking and polyethylene to create a tight seal.

Ventilation

Attic ventilation serves a number of purposes: it reduces summer heat build-up; after air sealing it is your second line of defense against water vapor that may have found its way into the attic; and finally, a colder, well vented attic space will be less prone to the formation of ice dams at the eaves.

Make sure that the existing attic vents are working properly, and are not blocked by insulation, dirt or other materials. You may have to locate roof or soffit vents from outside if they are not clearly visible from inside the attic.

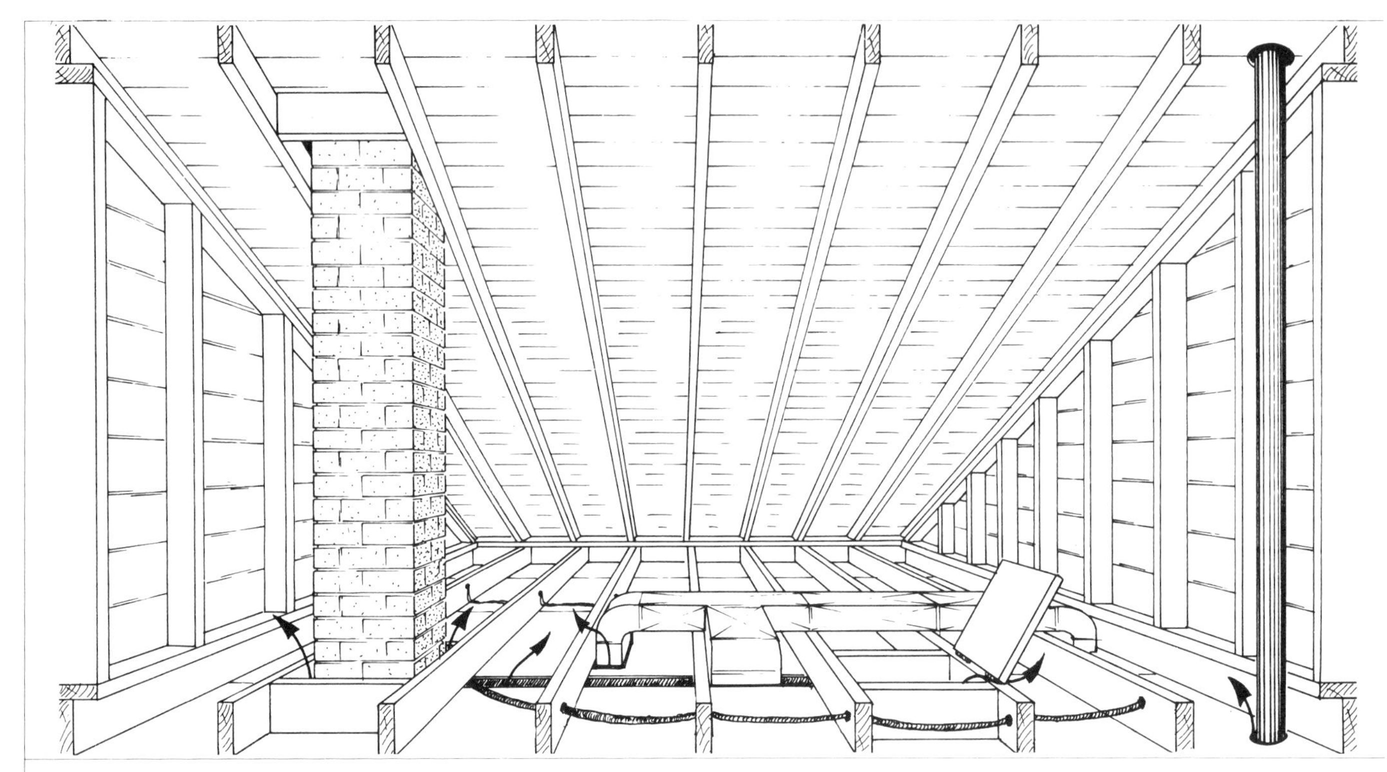

Attics have many potential air leakage paths.

- **Around attic hatches**. Attic hatches are an obvious, but frequently overlooked source of air leakage and heat loss. See page 41 for details.
- **Around the chimney**. The National Building Code requires that air spaces between chimneys and floor or ceiling assemblies through which they pass be sealed with a non-combustible firestop. See page 40 for details.
- **Along the edge of shared walls**. There is often a gap between the party wall and the edge of the attic floor. Ensure that this gap is well plugged.

Although an air-tight ceiling will significantly reduce the likelihood of moisture in the attic, building codes still require minimum attic ventilation. The ratio of vent area to ceiling area should be approximately 1 to 300. **Do not automatically increase ventilation**. Creating a tight ceiling is a far more effective way of eliminating moisture problems.

Electric exhaust fans are not recommended for attic ventilation. An electric exhaust can draw more air than can be supplied through the soffit vents. This will actually pull house air into the attic, resulting in greater heat loss and moisture accumulation.

The location of vents is as important as their number and type. The following sections will detail the best approach depending on your attic type.

Remember, vents are important but won't prevent condensation on their own. Ventilation alone will not solve the problems created by air leakage. Air sealing is your first line of defense.

Part II Easily Accessible Attics

After you have inspected the attic and carried out any remedial work, focus attention on air and moisture control, insulation and ventilation.

Air and Moisture Control

There are three options for installing an air barrier system in an unfinished attic:

- concentrating on air sealing;
- installing polyethylene over the joists; or
- installing polyethylene between the joists.

Of the three, the first is the most practical, since the installation of a polyethylene air and vapor barrier in an existing attic is fraught with obstructions and requires painstaking attention to detail.

If the attic retrofit is being completed in conjunction with interior renovations, the easiest approach is to install the new air barrier on the underside of the ceiling joists. Part VI deals with two approaches to this technique.

Concentrating on Air Sealing

Where and how to air seal has been covered in Chapter 3, and in Part II of this chapter.

In conjunction with a comprehensive air sealing job, it is a good idea to paint the room side of the ceiling with two coats of oil-based paint or a single coat of latex vapor-barrier paint to further inhibit vapor movement. Remember, however, that ordinary latex paint is not a suitable vapor barrier.

Check the attic during or just after a cold snap in the winter months. Some frost build-up is to be expected, but if it is particularly heavy, check to make sure that the ventilation is adequate. **And work harder at sealing air leaks into the attic!**

Installing Sheet Material Over the Joists

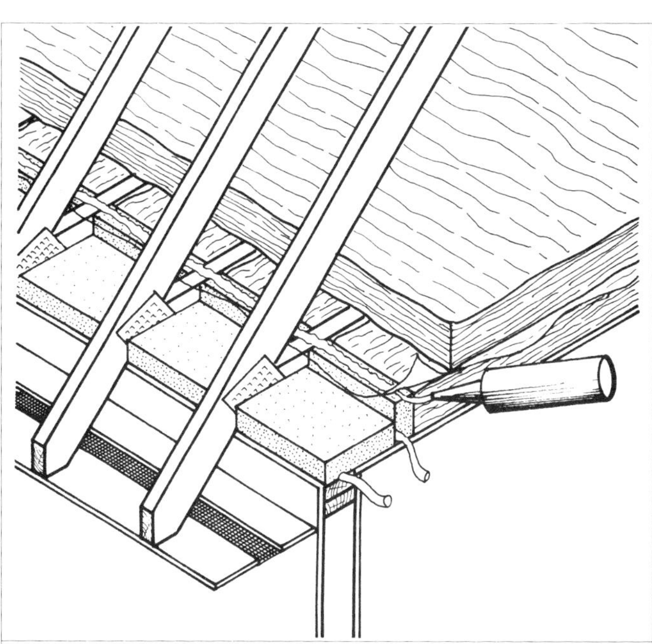

Installing polyethylene sheets over the joists, showing the eave detail using rigid foam boards.

If the attic is unobstructed with chimneys, plumbing stacks or structural members, install polyethylene directly over the existing ceiling joists. This method involves the least number of seams and requires less caulking and stapling than other methods. It also allows you to leave existing insulation in place. Seal all the obvious air leakage paths before laying down the polyethylene.

To avoid trapping moisture between the plastic and the wood – leading to possible wood rot or other moisture related problems – **install at least twice the insulating value** over top of the air-vapor barrier (the 1/3 - 2/3 rule). This means, for example, if the joist height is 89 mm (3 1/2 in.) and contains RSI 2.1 (R12), at least RSI 4.2 (R24) must be installed over top of the barrier.

The main difficulty with this technique involves sealing the barrier to the wall top plate, especially at the eaves where there is little room to manoeuvre.

Yet it is essential to seal this area as well. Rigid board insulation can help to bridge the gap in this area. Cut rigid board to fit between the ceiling joists and to extend from the exterior wall top plate towards the attic. A second piece of rigid insulation, installed vertically, joins the polyethylene to the horizontal rigid board. Carefully caulk any joints or seams between materials. Expanding foams are also useful for sealing areas around joists and boards.

Installing Sheet Material Between the Joists

Where obstructions make the previous method too difficult – e.g, a truss roof – install the polyethylene air barrier between the joists. This is a lengthy, painstaking process, so make sure you have plenty of patience.

If there is no insulation in the attic, your job is simplified. If there is insulation in the attic, it will have to be removed from the area you are working on and set to one side. Cut strips of the polyethylene about 200 mm (8 in.) wider than the joist spacing. Lay a bead of caulking on the side of the joists all along their length, and install the polyethylene, holding it in place with a series of staples (see illustration).

Laying polyethylene strips between the joists as an air and vapor barrier.

Remember, any obstructions in the attic, such as electrical wires or pipes, will require cuts in the barrier. These will need to be carefully sealed to make the barrier continuous.

Installing Insulation

The most common materials for use in an accessible attic are batt/blanket types or loose fill insulation. In some circumstances it may be a good idea to use a combination of types. If there are a lot of obstructions above the joists, such as with a truss roof, it may be easiest to put batt insulation into the joist spaces and then use loose fill to create a complete blanket of insulation above the joists and around all the obstructions.

On the other hand, if some spaces are irregular or obstructed, it may be easiest to use loose fill. You will have to choose the insulation types most appropriate to your situation — refer to the **Insulation Summary,** Chapter 2, for guidance.

Batt or Blanket Insulation

Batt insulation is simply pressed into place between the ceiling joists. If you purchased the correct width it will fit snugly. No stapling is necessary. Some other installation tips are outlined below.

- Butt the ends of batts together as snugly as possible.

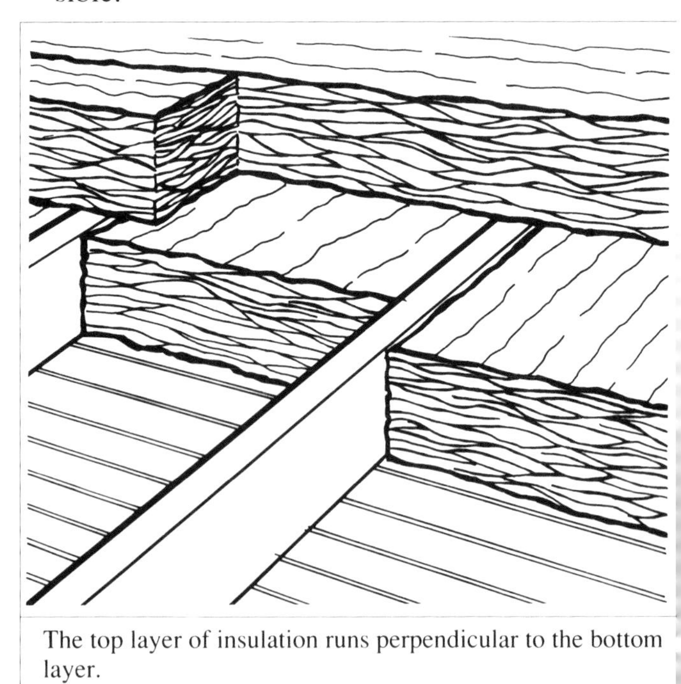

The top layer of insulation runs perpendicular to the bottom layer.

- **Batts should cover the top plate of the exterior wall but not block the venting**. To maintain air flow, leave a space of 37 mm to 50 mm (1 1/2 in. to 2 in.) between the top of the insulation and the underside of the roof sheathing. To prevent this space from being blocked, use baffles between each rafter space.

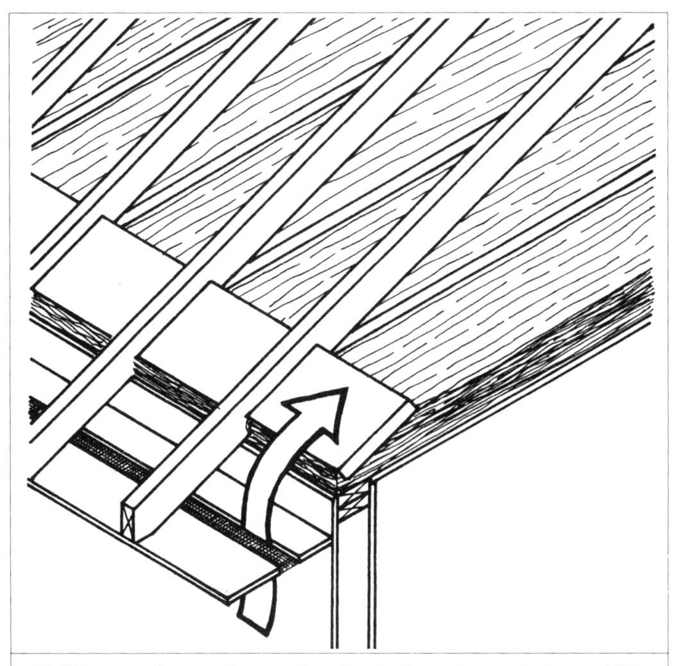

Baffles can be used to maintain air flow through the soffit vents.

- Insulate snugly around cross bracing using diagonal cutting as illustrated (avoid cutting the air barrier). Alternatively, you can cut one batt into a series of wedges, and then fit a wedge under each brace.

Fitting insulation around cross bracing.

- The first layer of batts should be thick enough to completely fill to the top of the joist space. The second layer can then run in the opposite direction, across the joists, blocking any heat flow through and around the joists (see illustration). Ensure that there is no gap between the two layers of insulation.
- Fill any awkward spaces or gaps with pieces of batts, or with loose fill insulation.
- Blanket insulation is applied in basically the same way as batts. It may be precut with a knife, or cut on the spot. Start at one end of the attic and simply unroll the blanket.

Loose Fill Insulation

Loose fill insulation can be poured by hand or blown in either by the homeowner with rented equipment (in which case only cellulose can be used), or by a qualified contractor.

- Loose fill insulation can be poured on top of the air barrier and vapor barrier. Level it with a board or garden rake, as illustrated. It's best to add insulation to a depth greater than the height of the joists. This extra thickness makes levelling a bit difficult but is worth it. Nail strips of wood to the side of some of the joists to help you gauge the depth of the insulation. **Maintain an even depth throughout the attic.**

Installing loose fill insulation. Note markers indicating final depth of insulation.

- Pour the insulation to fill all nooks and crannies.
- At the eaves, take care to keep the insulation from blocking the ventilation or from disappearing into the eave space. A piece of rigid board insulation, or a wood baffle should be installed before the work begins. Some building supply stores now sell cardboard or foam plastic baffles that can be stapled between the rafters. In any case, be sure that the insulation extends out far enough to **cover the top of the exterior wall**.
- If your loose fill is deeper than the joists, build a crib around the attic hatch so that it can be filled to the edge.

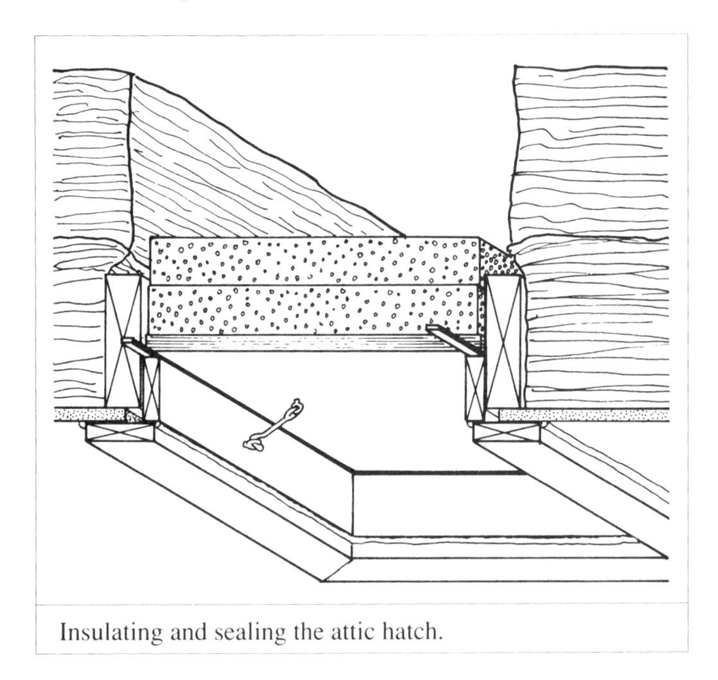

Insulating and sealing the attic hatch.

- The bags of material will list how many square metres (or square feet) each bag should fill in order to give the required RSI value. Knowing the size of the attic will help you determine the number of bags you will need.
- If you are having a contractor do the work, calculate the RSI value that you want and check the bags of insulation to be used. They should indicate what area one bag will cover at the selected insulating value. You and the contractor should then agree on the total number of bags to be used, the expected insulating value, and the *minimum* settled depth of insulation *throughout* the attic.

Additional Tips

- Fill the space between a masonry chimney and the wood frame around it with **non-combustible** insulation which must be certified to the appropriate CSA standard. Make sure that this is an application that is specified by the manufacturer. Leave 50 mm (2 in.) of space around factory-built metal chimneys to prevent any build-up of heat which might create a fire hazard. This space should be sealed against air leakage (see page 40 for details).
- If you live in a row house, and share a concrete block wall with your neighbour, this wall should also be insulated; otherwise the convection currents that circulate in the hollow core of the wall will transfer heat to your attic. (See Section III for details on complications in attics.)
- **Do not cover recessed light fixtures** and be careful not to insulate too closely around flue pipes and gas vents. Insulation will tend to cause a heat build-up and create a potential fire hazard.

Ventilation

Houses with peaked roofs and accessible attic spaces are the simplest to vent. The amount of attic ventilation is directly related to the size of the ceiling area in the building. In most cases, the ratio of unobstructed, free ventilation area to ceiling area should be approximately 1 to 300. Because these ratios refer to unobstructed ventilation area, the area must be increased because the vents are covered with **screening** (to keep out insects etc.) and with **baffles** (to keep out rain and snow).

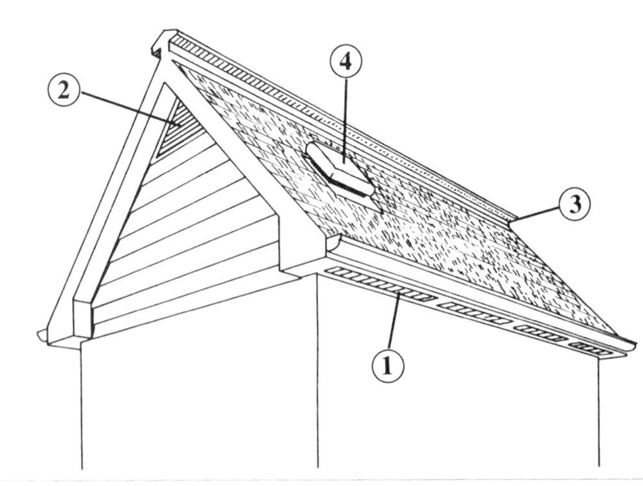

Different types of ventilation: 1) soffit vents; 2) gable end vent; 3) ridge vent; 4) roof vent.

Ideally, locate vents to allow for good cross ventilation from end to end and from top to bottom. This means placing vents at the eaves and at the peak. The drawing shows three types of vents – on the ridge, at the gable end, and near the peak of the roof.

Any of these are adequate when used in conjunction with under-eave (soffit) vents.

Fifty per cent of the ventilation area should be continuous soffit vents and the other 50 per cent gable, ridge, or roof vents. Ridge vents are preferable where practical, but they must be equipped with baffles to deflect wind blowing up the roof and to prevent the penetration of water and snow. Occasionally a house will have been built with no airflow between the soffits and the attic space. In this case install vents at opposite ends of the attic to take advantage of cross ventilation.

Part III The 1 1/2 or 2 1/2 Storey House

These types of attics have several small sections that sometimes makes access and insulating difficult. If you can't get into these spaces to work, you may have to hire a contractor. If the space is accessible, the following section gives some guidelines on how to do the work yourself.

The wall and floor section of a 1 1/2 and 2 1/2 storey house should be insulated.

Air and Vapor Control

Where you have access to the attic space, follow the directions for the control of air and vapor flows as outlined in the previous section. **Concentrate on sealing all air leakage paths into the attic**.

One major source of air leakage that cannot be overlooked is through the ceiling joists immediately beneath the knee walls. Prevent air leakage in this area by filling the space with a rigid insulation board, installed flush with the back of the knee wall finish. Seal the edges of the insulation as thoroughly as possible. Spray polyurethane may also be practical for controlling air flow in this space.

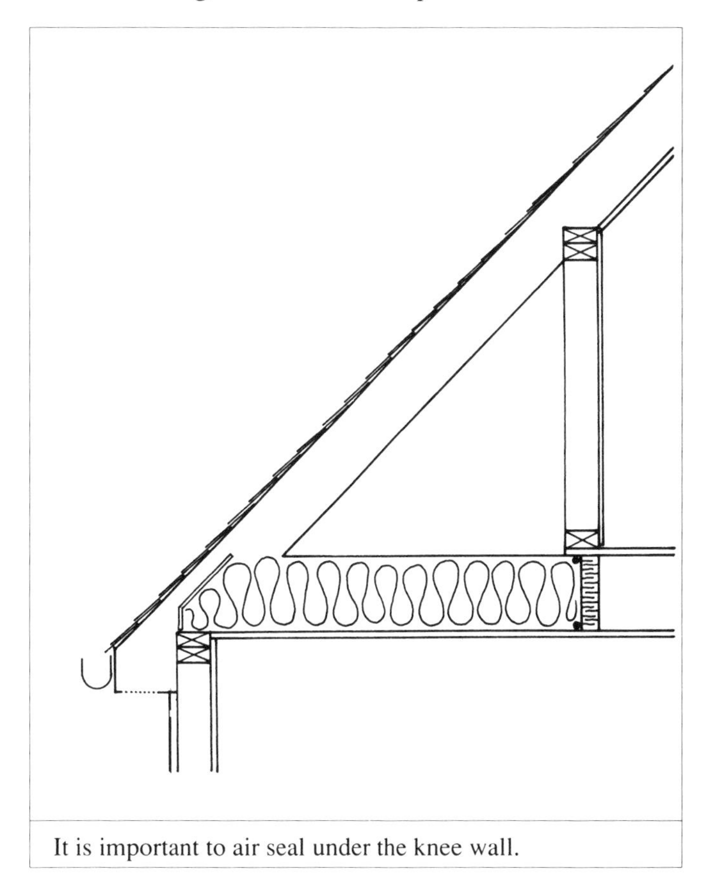

It is important to air seal under the knee wall.

Ventilation

Ventilate the attic spaces above and beside the top story separately, using gable vents. Make sure that the vents prevent wind from blowing through the insulation.

A second option is to use gable vents in the area above the attic ceiling and eave vents in the side areas. This is a good option if the house already has working eave vents. A ventilation space in the **rafter section** will permit the air to flow between the insulated areas. Otherwise, you will have to install air **channels** in the section along the rafters to ensure adequate ventilation (see diagram). These channels must be on **the cold side** of the space, which can then be filled with insulation.

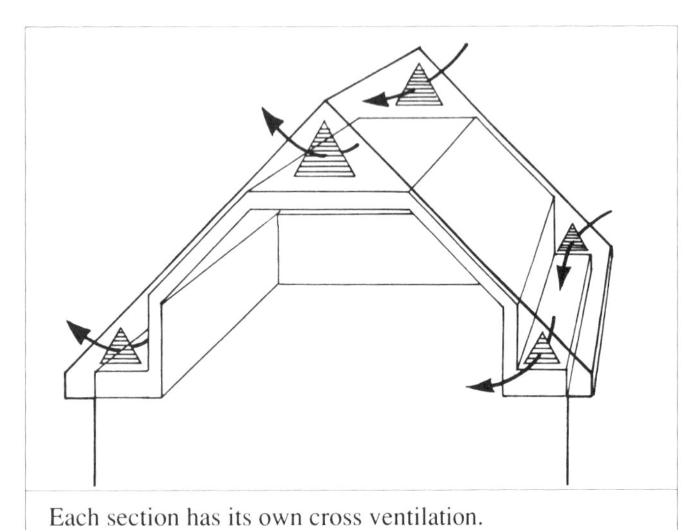

Each section has its own cross ventilation.

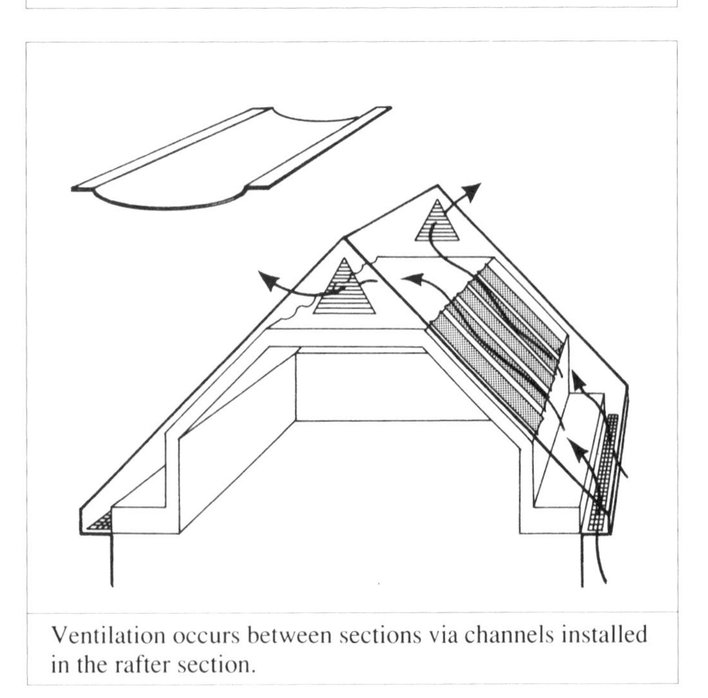

Ventilation occurs between sections via channels installed in the rafter section.

Insulation

Theoretically, a rigid foam insulation could be used; however, batts are less expensive and easier to handle in confined spaces. If there are any electrical outlets (careful!) or pipes in the knee wall, make sure that you keep them on the **warm** side of the air and vapor barrier and insulation, and seal the air and vapor barrier around them.

- Treat the **outer attic floor and the attic space over the 1/2 storey ceiling** (see above figure) exactly as described for standard, unfinished attics on the previous pages.
- The **end walls** are the full-height walls that are exposed to the exterior. Treat these exactly as described in Chapter 6 by blowing insulation.
- The **knee wall** can be treated like an unfinished attic floor (see page 47), making allowances for the fact that it stands upright. Install a combination air and vapor barrier made from polyethylene strips sealed between each stud. The air barrier can also be created by sealing all cracks and penetrations, and a vapor barrier painted on the interior surface. Next, install the insulation and secure it in place. This can be done with friction fit batts, or by securing the insulation with building paper, cardboard, olefin sheets or string or wire. Rigid board insulation nailed to the exterior side of the studs can also be used to hold the batt insulation in place. This will increase the thermal resistance of the wall section and reduce thermal bridging.

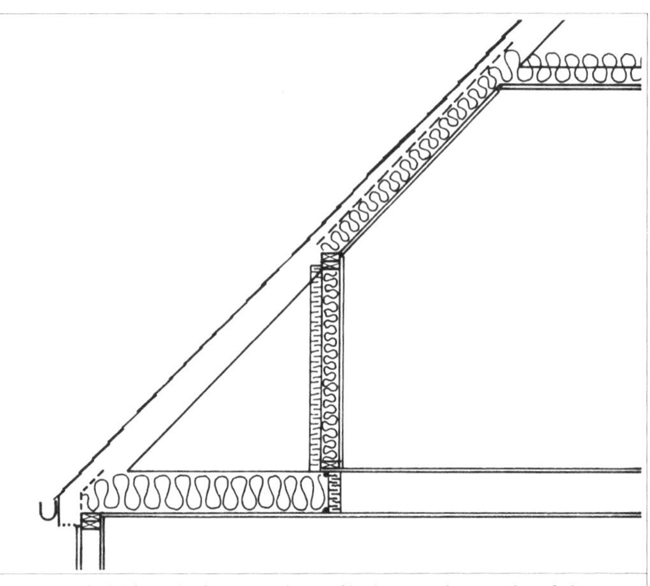

Extra rigid insulation can be nailed over the studs of the knee wall section.

- The **section in the rafters** may be filled with insulation if all penetrations through the ceiling are sealed, and if this is permitted by local building codes or standards.

Other Complications

Wall of Heated Room

Some houses have a wall in the attic that adjoins a heated space. Insulate it as you would a knee wall. See pages 51 and 52 for instructions.

Shared Wall

Semi-detached or row houses that share a concrete block wall will lose heat into the attic because concrete is a good conductor of heat, and air circulates inside (and through) this wall. Ideally, the shared wall should be plugged at the ceiling level by having a contractor drill holes and inject small amounts of polyurethane foam into the blocks. In many cases, this will not be possible or economical. If there is a wood frame party wall at the top of the block wall, air seal the junction at the top of the block wall.

The next best alternative is to insulate both sides of the exterior surface of the shared wall in the attic. Paint the wall first with an impermeable concrete paint or cover it with polyethylene. Next, tightly secure a layer of insulation to the wall. The wall normally goes through the roof line and is plugged. However, if there is space at the top of the wall between the concrete blocks and the roof, or any gaps, they should be plugged and covered with insulation material.

Dormer Windows

Many 1 1/2 or 2 1/2 storey houses also have dormer windows. The walls of the dormer may be insulated with batts, as described for the knee wall. Remember that the air and vapor barrier should be on the warm side of the insulation, and sealed at all joints and corners.

The remaining walls and the dormer ceiling are much more difficult. The simplest solution would be to have insulation blown-in. Remember to seal any ceiling fixtures or other penetrations.

Floorboards

Some houses will have the attic floored over, even when it's not used as living space. You can insulate it by lifting the floorboards and treating it as you would an unfinished attic. Some or all of the boards may have to be replaced to maintain the stiffness of the ceiling.

Alternatively, a contractor can fill the subfloor space completely by blowing in loose fill insulation through access holes. Air leaks, such as around plumbing stacks, should be sealed first. To ensure that the space gets filled completely and at the right density, you and the contractor should agree on the insulating value to be achieved and the total number of bags of insulation to be used.

Before insulating, either you or the contractor should check for damaged or frayed wiring or recessed light fixtures or other sources of heat that may be concealed beneath the floorboards. All heat sources must be protected from the insulation or removed entirely.

You will not likely be able to achieve the minimum recommended levels for attic insulation by filling this space alone. It might be worthwhile to add some insulation over the floorboards to keep them warm and reduce thermal bridging through the joists.

Part IV Special Cases

If the attic is both cramped and without a hatch, all is not lost. You may be able to cut a hole in the ceiling in an out-of-the-way place, such as a closet, or you may be able to gain access through outside vents, either existing or new ones. So don't give up, read on!

If you discover that the attic is cramped, but large enough to work in, follow the instructions in Part II. If, on the other hand, there really is no free space at all above the ceiling joists – continue with this section.

How to Insulate an Attic That Is Too Cramped to Work in.

Basically you have two choices. It may be possible to insulate outside the building on top of the existing roof (see Part V), or it may be possible to have a contractor blow in loose fill insulation. Choose the contractor with care — make sure that the firm has experience in this type of situation.

If you choose to have insulation blown in, calculate the RSI value that you expect to achieve and check the bags of insulation to be used. They should indicate what area one bag will cover at the selected RSI value. You and the contractor should then agree on the total number of bags, the expected RSI value, and the **minimum** depth of insulation to be achieved **throughout** the attic.

There will likely be no way of installing a new sheet material air barrier. If one does not already exist, it should still be acceptable to install insulation **if the following conditions are met** (even if a barrier is present, these points are worthwhile):

- there is no evidence of moisture problems;
- humidity levels in the house are reasonable (see Chapter 1); and
- any air leaks through the ceiling into the attic are sealed.

You can achieve added protection by painting the ceiling below the attic with a coat of latex vapor barrier paint or two coats of oil-based paint.

Ensure that the contractor makes provision to stop insulation from entering the eave and blocking the ventilation. If there are any recessed light fixtures or other sources of heat in the attic, make sure precautions are taken to avoid creating a fire hazard.

Finally, if there are any vertical walls in this attic space (e.g., knee walls), they likely cannot be insulated by blowing in insulation. Consult your contractor for details.

Ventilating these attics can be difficult because of the limited space and the difficulty of creating an adequate airflow. Where the roof extends over the exterior walls it may be possible to use soffit vents, in combination with roof vents or build up ridge vents.

Many houses with cramped attic spaces lack eaves. In such cases, approach ventilation with caution. The best approach is to carefully seal the ceiling below the attic from inside the house, and then insulate without installing additional vents.

If possible, check the roof space for moisture problems during or just after a cold snap in January or February. Some frost is to be expected, but if the build-up is especially heavy, you will have to consider ventilating the space and work even harder at locating and sealing all the air leaks and reducing humidity levels in the house.

In any event, it is advisable to check with local building authorities to determine what procedures are permitted in your area.

Cathedral Ceilings and Flat Roofs

A house (or any portion of a house) with a flat roof, cathedral ceiling or some other "atticless" construction is likely to be a difficult case, and will require the services of a qualified contractor.

The main problem with these roofs is the limited space for insulation and ventilation. In fact, if there is already some insulation in the joist space, adding more may simply not be economical.

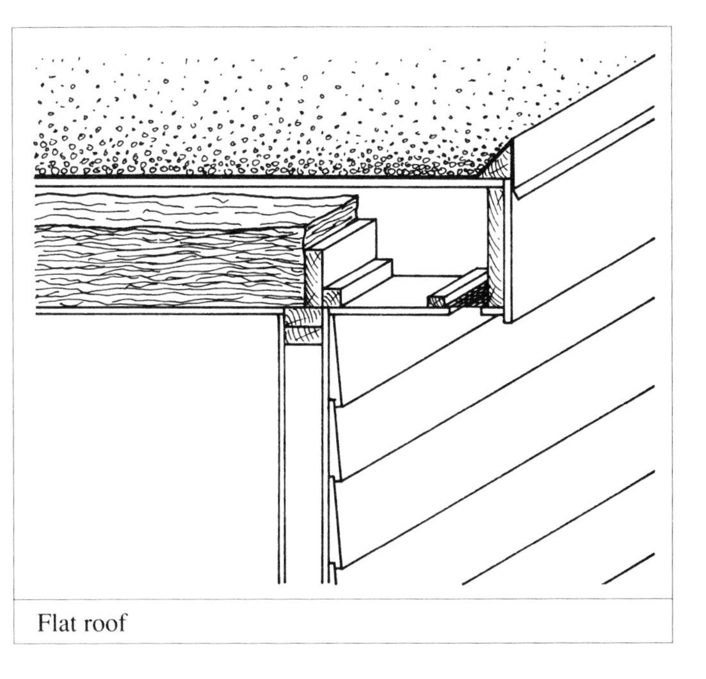

Flat roof

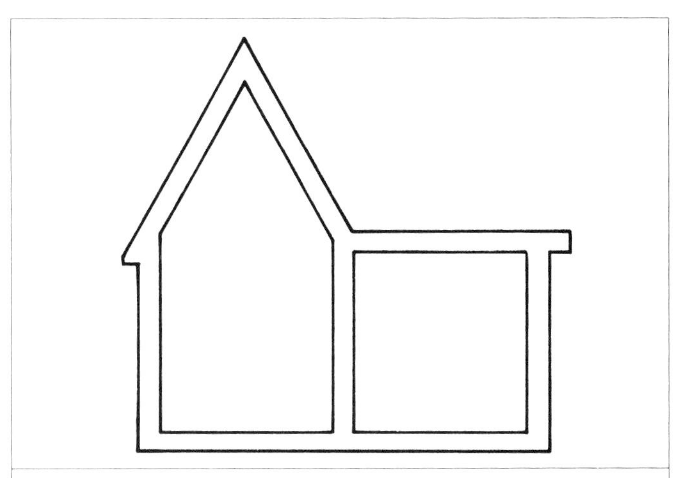

Cathedral ceilings and flat roofs are difficult to insulate and maintain a ventilation space.

However, if you decide it is worth your while to increase insulation levels, there are a number of options. **Each option involves some risk of either moisture problems or thermal bridges that can reduce the effectiveness of the insulation.** A technique involving blown insulation is discussed below. Section V discusses both interior and exterior retrofits, including the addition of a new roof.

- The existing space between the ceiling and roof can be blown full of loose fill insulation by a contractor. Since this eliminates ventilation **it is not a generally recommended practice**. Take extra care to make sure that air leaks into the ceiling are sealed from below. This is very difficult because wiring and plumbing usually puncture the ceiling in a number of places. Moreover, the partition walls may not be completely blocked off at the top, allowing large amounts of air to flow through the interior walls into the ceiling. Where the interior walls are completely open to the ceiling, there is no easy solution – unless you are prepared to have your interior walls blown full of insulation and sealed along all trim, outlets, and other penetrations.

  If you choose to fill the interior of a flat roof or cathedral ceiling, the most appropriate material is probably **cellulose fibre** blown in to a high density (56 to 72 kg/m^3 or 3.5 to 4.5 lb./ft.^{3}). The contractor should calculate and confirm the density for each roof cavity. The high density of the insulation – combined with comprehensive air sealing – should reduce airflow sufficiently to avoid condensation problems.

Part V Renovation Activities

Renovations or repairs provide an opportunity to ensure a well-insulated attic ceiling and walls with an effective air and vapor barrier.

Attics to Be Finished and Heated

A popular renovation activity is to convert an existing unfinished attic to new living space. This can be as simple as insulating and drywalling the existing space. It can also be as complicated as adding new floor joists, lifting part of the roof for more headroom, and installing a stairway.

Depending on the depth of the rafter space, it may also be difficult to achieve the high RSI values recommended for attic insulation. If you choose to do it, take the following steps:

- Install **collar beams** (collar ties) between every pair of rafters, as illustrated. These will provide structural support for both the roof and the new ceiling.

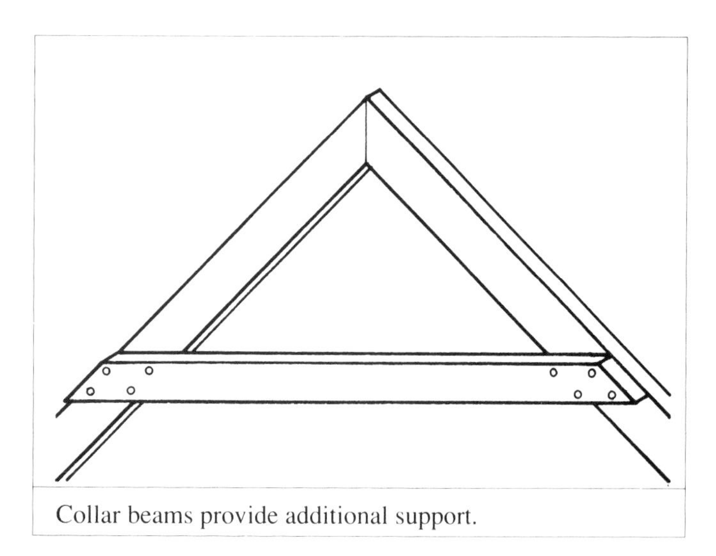
Collar beams provide additional support.

- Ensure that the insulation to be installed in the rafter spaces will fit snugly against the top of the walls to form a continuous thermal envelope. There must be no gaps around the perimeter of the attic floor or heat will escape and ice damming may occur near the eaves. In most cases, it should be possible to insulate each rafter space all the way down to the eave – over the top of the exterior wall.

  When insulating each rafter space, leave a clear space between the top of the insulation and the underside of the roof sheathing to allow for roof ventilation. Check for local code requirements.

  Warm air under the attic floorboards will still be able to bypass this insulation. To prevent this, you will have to block and seal the joist spaces along the perimeter of the attic floor. If the floorboards can be temporarily lifted, fill the space between the joists with a piece of low-permeability rigid board insulation and seal it to the joists. If the floorboards are not easily lifted, it may be possible to spray polyurethane between the joists along the perimeter from the rafter space beyond the floorboards.

- Finally, staple a **continuous** polyethylene air and vapor barrier to the rafters and end wall studs, taking care to seal the edges and seams with acoustical sealant.

- If your attic room is not too cramped, consider building onto the rafters and studs to allow for more insulation (up to the recommended minimum level for attic insulation, if possible). After installing insulation in the existing spaces, nail 38 mm x 38 mm (2x2) strapping perpendicular to the rafters and studs. Space the strapping to suit the width of the insulation you will use. However, the strapping should be no more than 600 mm (24 in.) apart — measured on centre — or you will have difficulty attaching the interior finish. Fit insulation snugly between the strapping covering all the rafters and studs. Finally, staple a **continuous**, sealed polyethylene air and vapor barrier to the strapping.

- Insulation between the collar beams is applied from below in much the same way, with a **continuous** polyethylene air and vapor barrier applied last. If the collar beams have already been insulated, and if there is access to the upper portion, then more insulation may be added as in a normal attic.

- Finally, block off and seal any vents into those parts of your attic which are now heated.

Dropped Ceiling

Where you have the headroom, constructing a dropped ceiling to hold insulation is an excellent way of thermally upgrading a cathedral ceiling or flat roof, especially when planned as part of a renovation. Several options are outlined below.

- Construct a new ceiling immediately below the existing ceiling. If the roof has exposed joists or beams, usually for decorative purposes, it may be possible to close the space in, creating a new ceiling. Batts, or a rigid insulation can be installed in the space followed by a continuous air and vapor barrier and the new ceiling. In all cases, moisture could cause difficulties. Review the section on moisture control in Chapter 1 before considering this option.

  If you choose to build a new ceiling from below, remember that you have to prevent warm air from getting into the new cavity space and bypassing the insulation. This will involve sealing the perimeter of the new ceiling as well as any possible air-leakage paths through partition walls.

- Existing rafters can be extended to accommodate additional insulation. This can be done by cross-strapping the existing rafters (see diagram), or by extending the rafter cavity with 38 mm x 89 mm (2x4) lumber and plywood gussets.

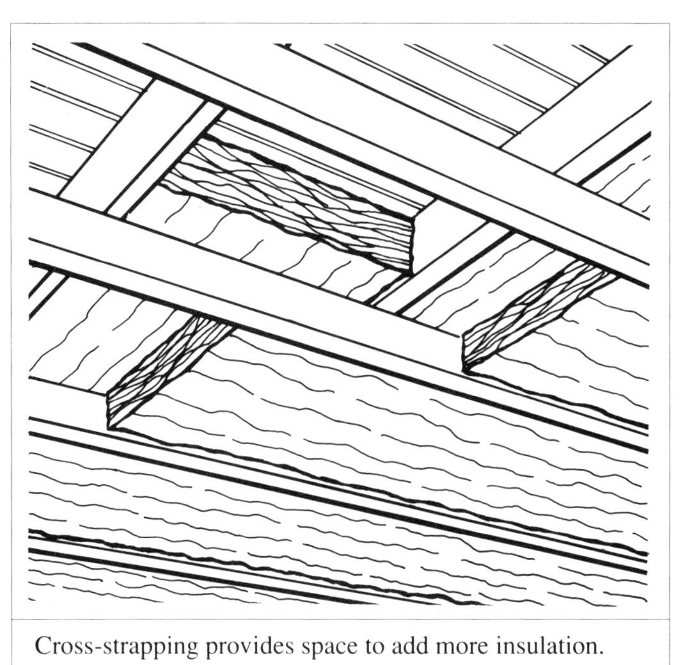

Cross-strapping provides space to add more insulation.

Extending the rafters provides space for insulation and ventilation.

Though it is not necessary to remove the interior finish and expose the rafters when cross-strapping — as it is when extending the rafters — it is a good idea, since you can check the state of the insulation and see if there is a vapor barrier.

- You can add rigid insulation directly to the underside of an existing ceiling. This technique allows the existing ceiling finish to be left intact, but does not allow access to the cavity.

Rigid board insulation must be mechanically fastened directly to the roof structure and with adequate nailing surfaces provided for the new ceiling finish. If the rigid board insulation is doubling as the air barrier, make sure the boards are tightly fitted and that seams are well sealed. Electrical fixtures will have to be extended to accommodate the depth of the new ceiling.

Adding a New Roof

Insulation can also be added on top of an existing roof. This option is suitable for cathedral ceilings but can really only be justified when major exterior alterations such as a new roof are required. For an average house the cost of reshingling, installing new sheathing, replacing eavestroughing, soffits and fascia, disposal and haulage costs, and the costs of installed insulation will approach several thousand dollars. You will almost certainly need the services of a qualified contractor.

- A relatively straightforward method involves installing rigid board insulation over the existing roof (see diagram). The higher RSI-value of rigid insulation means a smaller increase in the roof thickness, though several layers of insulation may be needed to meet the desired RSI level.

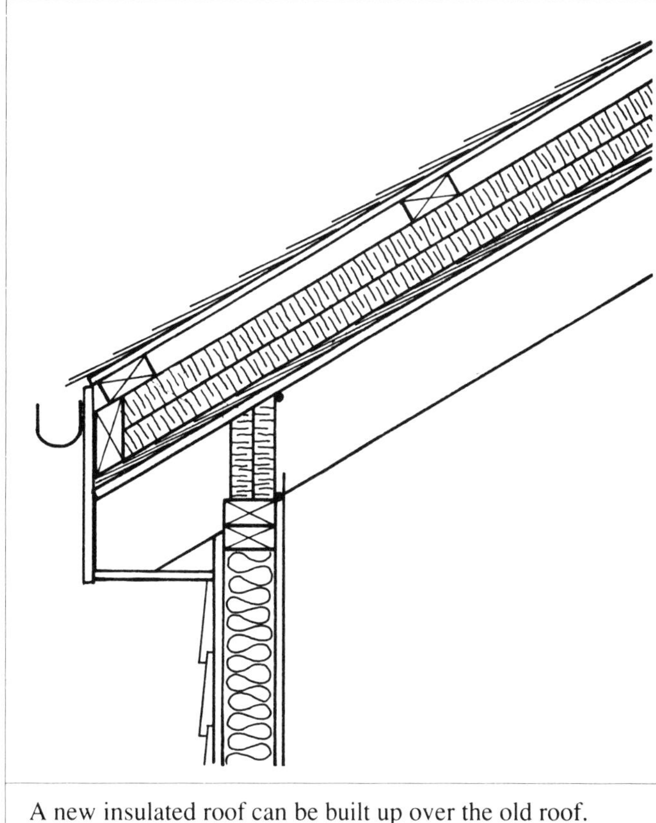

A new insulated roof can be built up over the old roof.

Sheet polyethylene is first placed over the existing roof to provide the air and vapor barrier (providing the 1/3 – 2/3 rule is followed). Or, low-permeability rigid board can be installed with the joints taped to prevent air leakage. This step is not necessary if there is already an air barrier.

It is very important to insulate and air-seal the spaces between the joists along the perimeter of the roof. This is to prevent heated air from escaping around the insulation. Install blocking between the joists. Line up the edges of the blocking with the inside finish of the exterior wall and thoroughly caulk all seams.

- A new roof can be framed over the existing roof and filled with batt insulation. The addition of a new roof frame may add to the structural loading of the entire assembly. Check with a building inspector before beginning the work.

Again, if required, a new air and vapor barrier can be provided by laying a sheet of polyethylene directly over top of the existing roof sheathing.

The load of the new roof must be transferred to the existing structure evenly and consistently. This is best done by installing cross members (purlins) spanning the length of the roof. Three purlins – located at the base, the mid-point and the roof peak – will generally suffice. The size of purlins will vary depending on the level of insulation to be installed.

Once the purlins are in place the new rafters are installed, followed by the insulation. Place the insulation so that it prevents air movement and thermal bridging.

Leave clear space between the top of the insulation and the top of the new roof rafter to allow for ventilation. Check your building code for details.

Refinishing the Interior

If the attic retrofit is part of interior renovations, consider removing the ceiling and installing a new sheet polyethylene or sealed drywall air barrier on the underside of the ceiling joists.

While sealing the air barrier on the ceiling to the one on the wall should pose no difficulties, maintaining continuity at interior partitions will require some ingenuity and detailed work. Where partition walls run perpendicular to the ceiling joists, maintain continuity by working from above, using connecting strips of polyethylene or extruded polystyrene. Where partition walls run parallel to the ceiling joists, install blocking and nailing strips to provide support for the new ceiling materials.

Sealing over partition walls.

It is also possible to leave the ceiling in place and apply a new layer of drywall, making sure that all penetrations are sealed. This method requires less labour and creates less of a mess. Create the vapor barrier by painting the ceiling with two coats of an oil-based paint, or a single coat of latex vapor-barrier paint.

New insulation is added in the attic according to the principles set out in Part II.

5 Basement Insulation

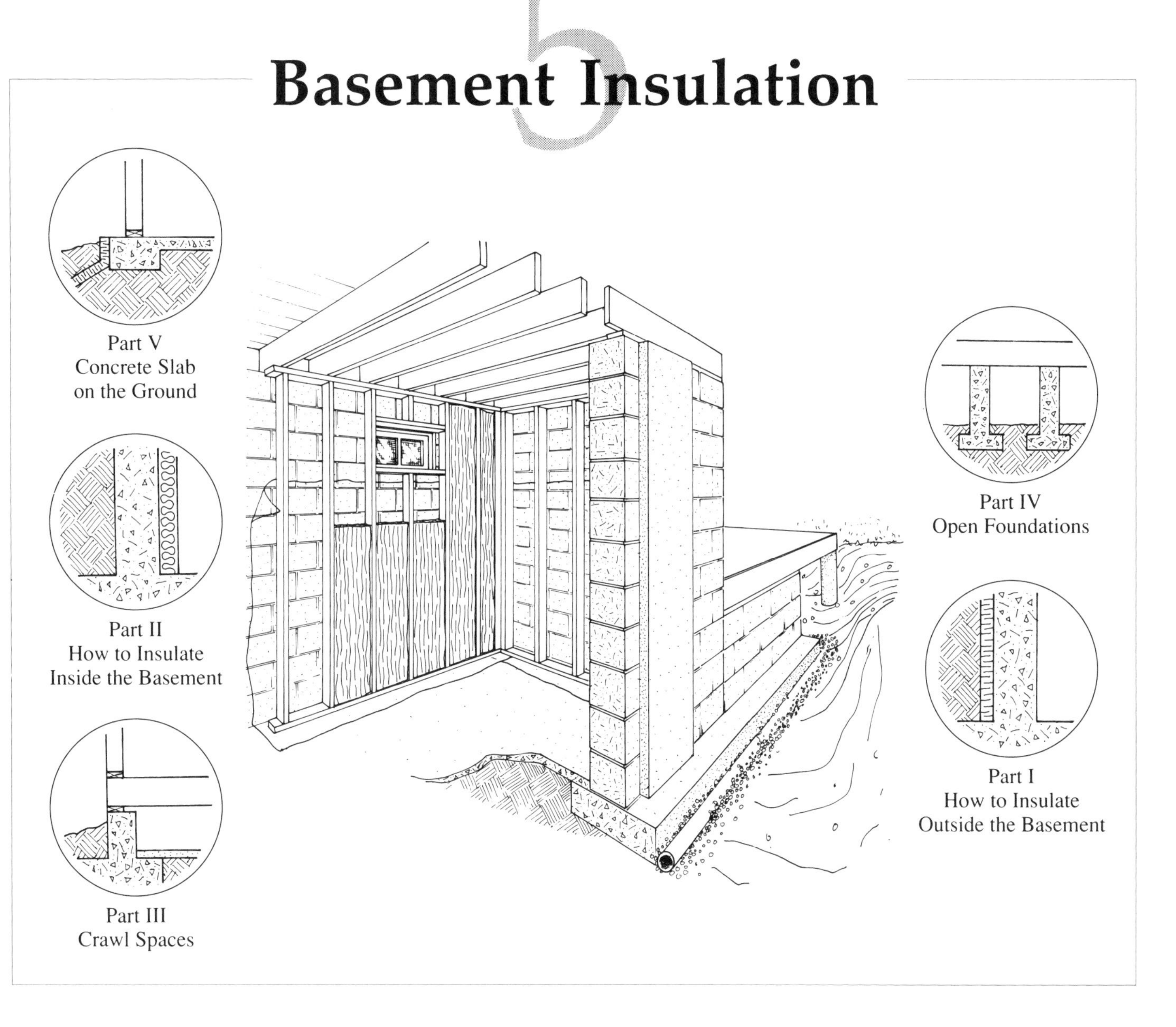

Introduction

How Important Is Basement Heat Loss?

Most homeowners don't think of their basement as a prime source of heat loss, yet basements can account for 20 to 35 per cent of the total. Basements lose so much heat because of the large, uninsulated surface area both above and below grade level. Contrary to popular opinion, earth is a poor insulator. There is also a lot of air leakage through basement windows and penetrations, through cracks, and at the top of the foundation wall (sill area).

Few basements have any insulation, and for most homeowners this means there is much potential for improvement. Insulation can often be tied in with other repair or renovation work, such as damp-proofing or finishing the basement.

Types of Basement Construction

The most common type of basement is the **full foundation** basement. It may or may not be finished and considered a living space. Full foundations can be made of many different materials, as described below.

Many houses have been built with a **partial depth foundation** that creates a crawl space under the house. Some older homes and cottages are built up on **posts and piers**. The space below the house is open to the outside, although this can be blocked off. Other houses are built on a **slab-on grade** where there is no basement or crawl space at all.

Whatever the type of construction, most basements/ foundations have no insulation at all. Even if your house is new, built since the mid-seventies, it may at best have partial-depth insulation.

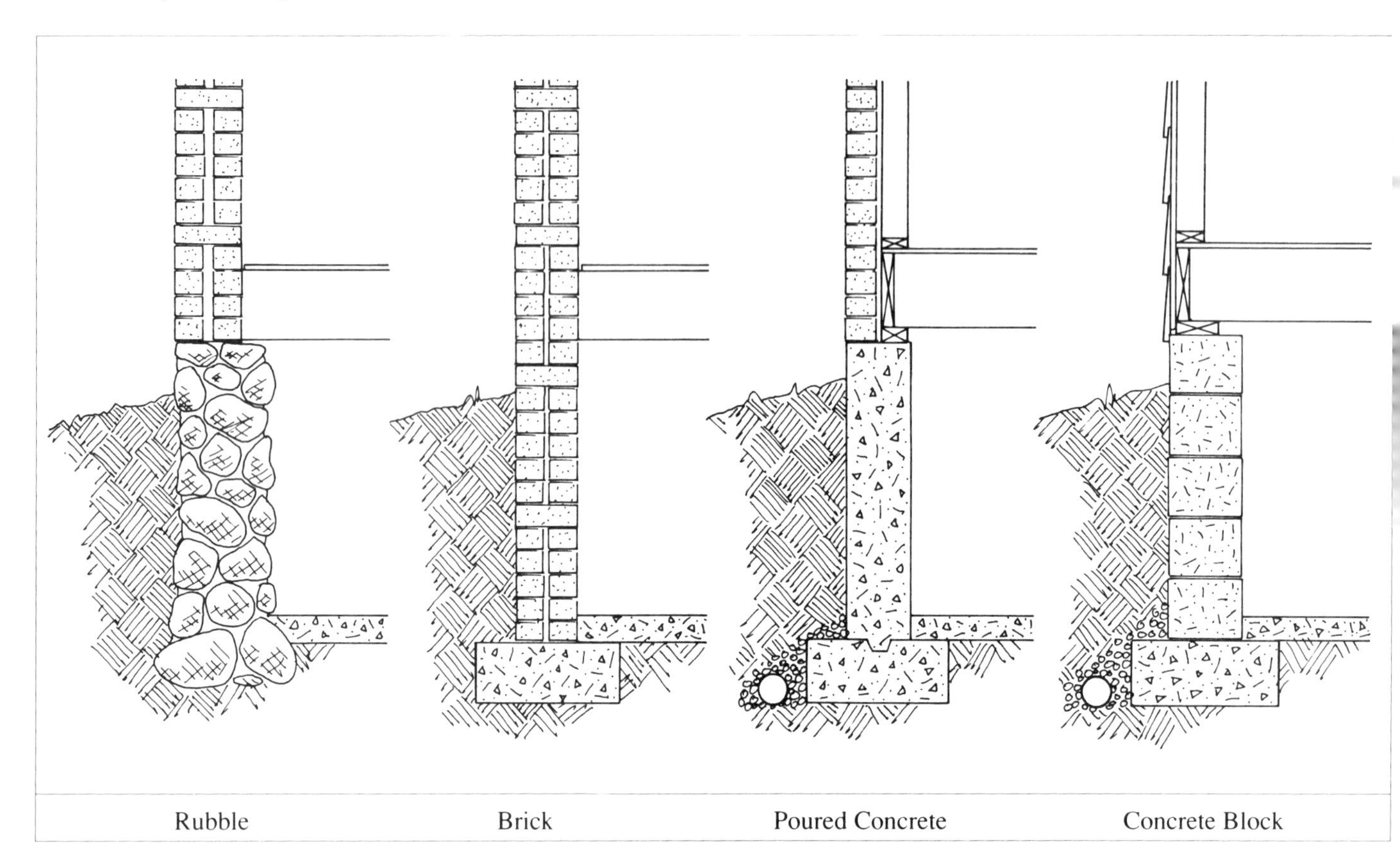

Rubble | Brick | Poured Concrete | Concrete Block

Concrete Foundations
Poured or concrete block foundations have been built since the 1920's, usually with parging, damp-proofing, and drain tiles on the exterior. This type of basement can be insulated from the outside or inside as long as there are no serious water or structural problems.

Older Foundations (rubble, brick, stone)
Older foundations made of rubble, stone or brick are often uneven and can vary in depth and thickness. These foundations were rarely damp-proofed, and have a high mortar content which can absorb water from the soil. They usually have a history of moisture problems and should be insulated from the outside.

Other Types of Foundations
Many newer homes are built with **preserved wood** foundations. These are made with specially treated wood studs and sheathing, and are generally fully insulated.

Many homes are built without a conventional basement. Examples of these include houses with **crawl spaces** and **slab-on-grade**. These types are covered at the end of this chapter.

Basement Assessment

Before planning the job take a good look at your basement. Correct any problems first. They will not go away just because you have hidden them behind insulation! Some of the problems to look for are noted below.

Water Leakage
Major water leaks (persistent leaks and flooding, in the spring and when it rains) must be corrected. This could be as simple as sloping the grade and directing downspouts away from the foundation. Often the solution requires excavating, damp-proofing, the addition of a drainage system, and insulating from the exterior.

Minor water leaks can sometimes be corrected by directing water away from the foundation and patching the foundation on the interior.

Correct problems with sump pumps or sewer backup before insulating.

Dampness

Symptoms of dampness include staining or mold growth, blistering and peeling paint, efflorescence (a whitish deposit on the surface), spalling (deterioration of the surface), as well as a musty smell. Minor dampness may be corrected from the interior, more serious problems should be corrected from the outside. Condensation can also form on the foundation walls in the summer because the air is very humid and the basement is cool.

Cracks

If it appears that you have an "active crack", that is one that gets bigger or smaller, you should seek professional help to determine if the situation requires structural repairs.

Which Approach: Inside or Outside?

In most cases, insulating on the outside is best from a technical point of view. Despite this, it is often necessary to insulate from the inside for economic and practical reasons. Sometimes a combination of approaches is required. Examine the advantages of each approach carefully and choose the one most suitable to your situation.

Insulating Inside

This may involve installing a wood frame wall and adding batt insulation (see illustration below). Another option uses rigid board insulation, with prefabricated metal channels or wood framing to hold the insulation, followed by a layer of fire resistant material (e.g. gypsum board), mechanically fastened to the wall. Normally, a moisture barrier is applied to the inside face of basement walls up to grade level, and an air and vapor barrier is installed on the warm side of the insulation.

The advantages of inside insulation are as follows.

- It can be incorporated into a program of finishing your basement.
- The work can be done any time of the year, and be done a section at a time.
- It is often an easier and cheaper way to insulate the full wall and to achieve high insulating values.
- The landscaping around your house, such as plantings and driveways, will not be disturbed.

The disadvantages of inside insulation are as follows.

- **Insulating from the inside should not be attempted in basements with a moisture problem.** If your basement has a history of damp or dripping walls you should insulate on the outside. If you must insulate on the inside, corrective measures are necessary to eliminate the moisture problem before adding insulation.

- Obstructions such as electrical panels, wiring, plumbing, stairs, partition walls, the oil tank, and so on, make the work more difficult and the insulation and air barrier less effective. If part of the basement wall is already finished, this too may prove troublesome (although wall panelling may be easy to remove and reinstall).

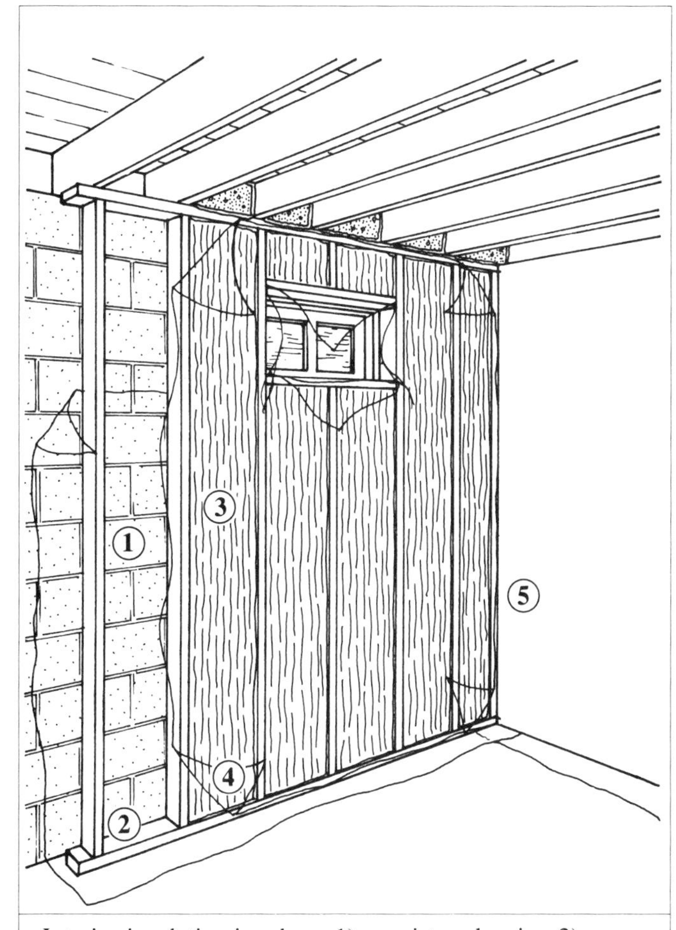

Interior insulation involves: 1) a moisture barrier; 2) new frame wall; 3) insulation; 4) air and vapor barrier; 5) finishing.

Foundations and Frost Heave

Some authorities have expressed a concern for the possibility of frost heave and structural damage when foundations are insulated from the inside. The concern is that frost will penetrate much deeper against the foundation. Extensive surveys and research have not found this to be a problem. Under some circumstances, such as particularly expansive clay soils in extreme climates, there could be a problem with some techniques. Check with your local building authorities, or see if your neighbors have experienced any trouble.

Insulating Outside

This involves excavating around the foundation, damp-proofing, and installing rigid insulation as shown in the illustration. Flashing must be attached to keep the water out from behind the insulation, and a protective covering must be installed on the exposed sections of insulation.

The advantages of outside insulation include the following.

- The outside wall tends to be more continuous and easier to insulate once the soil is removed.
- You can effectively correct any moisture problems. Rubble or brick foundations, and foundations with water leakage, dampness, or other moisture problems **must** all be insulated from the outside. Foundation repairs, damp-proofing, and the installation of a drainage system can be done at the same time.
- There is no disruption in the house and no inside space is lost.
- The mass of the foundation is within the insulated portion of the house and will tend to even out temperature fluctuations.

The disadvantages of outside insulation include the following.

- Difficulty might be encountered when digging a trench around the house. Excavating by hand can be a tedious, back-straining job. It's much easier if you have machinery do the work.
- Storing the dirt can be a problem.
- Excavation cannot be done in winter, and can be a problem in the spring or throughout the year if you are in a high water table situation.
- Features such as non-removable steps, paved carports, shrubbery, trees or fences can make the work difficult.

The following sections provide a step-by-step explanation of how to insulate outside and inside the basement.

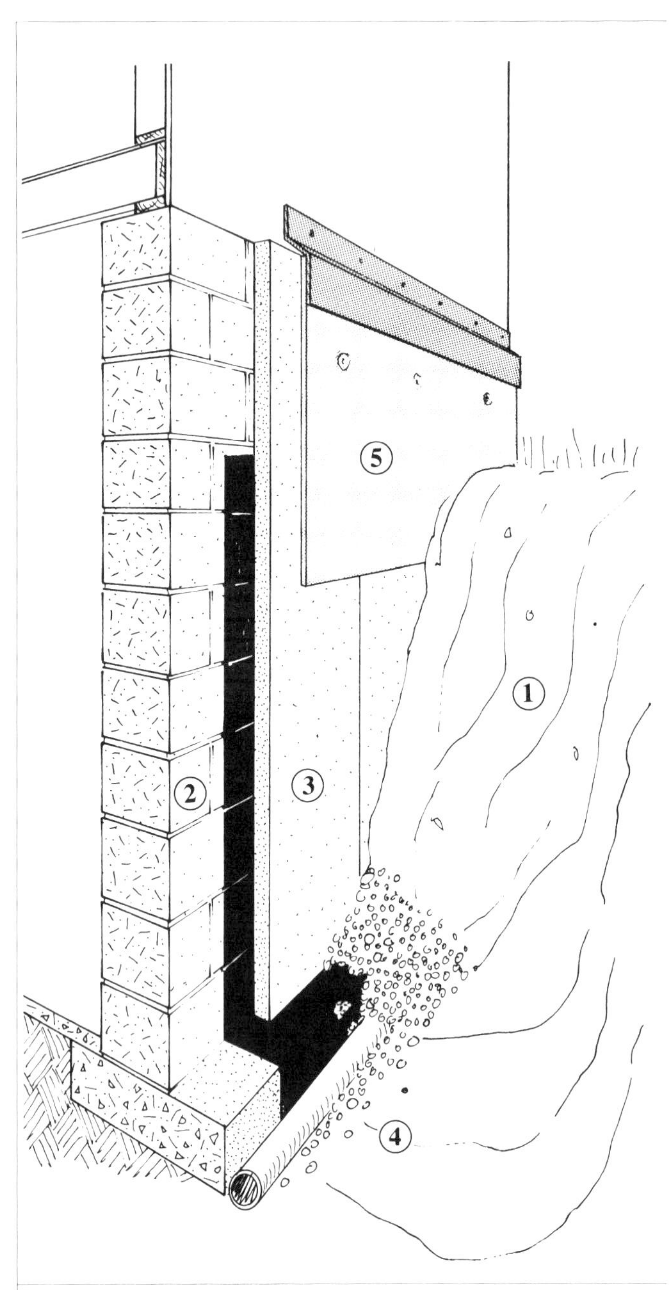

Exterior insulation involves: 1) excavating; 2) damp-proofing; 3) insulation; 4) drainage system and back-filling; 5) protective coating and flashing.

Part I How to Insulate Outside the Basement

Assessing the Situation

Whatever the basement type, there are a number of things to determine before beginning work:

- outside features that may inhibit excavation and insulating (porches, driveways, services, landscaping, access, lot lines);
- indications of structural problems (cracks, spalling, powdering mortar);
- signs of moisture problems (efflorescence);
- insulation requirements (type and thickness, height and depth);
- the site where the soil can be stored; and
- finishing details (protective coating, flashing).

Tools Required

- Tools for excavating (from a pick and shovel to a back-hoe) and installing insulation
- Tools for applying the flashing (depends upon your choice of flashing)
- Tools for adding batt insulation to the joist space from the interior (if necessary)

Special Safety Considerations

(Refer to Chapter 2, Part IV, for general advice on safe working procedures)

The job could involve extensive digging. Take it easy.

If any underground services enter your home (gas, hydro, telephone, water, sewage) find out where before digging. This is a free service offered by the utilities.

Some rubble foundations may be partially supported by the soil. If you suspect this may be the case, get some expert advice before you dig.

How to Insulate Outside the Basement — Step by Step

The work may require several weeks of effort. Plan for extra time if several steps are needed, such as excavating, repairing cracks, damp-proofing, installing a drainage system, etc.

1) Digging the Trench.

The excavation should be down to the footings. **Never dig below the bottom of the footings.** The width of the trench should give you room to work in. It's a big job! Don't make it bigger by oversizing your hole. You can dig by hand, or you may want to have it done by a contractor with the appropriate machinery. The excavated dirt can be stored on a tarp or sheet of polyethylene at least 600 mm (24 in.) away from the excavation. Protect the trench from rain, running water, the elements, and ensure that people or animals can't fall in.

Some soils are not stable and may require bracing to prevent them from falling in. Consult an expert if you are in doubt.

2) Preparing the Surface and Site

First clean the surface of the foundation with a wire brush and scraper. Inspect the foundation for major holes, cracks or damage. Repair where necessary. Smooth uneven surfaces with parging. Allow repairs to dry.

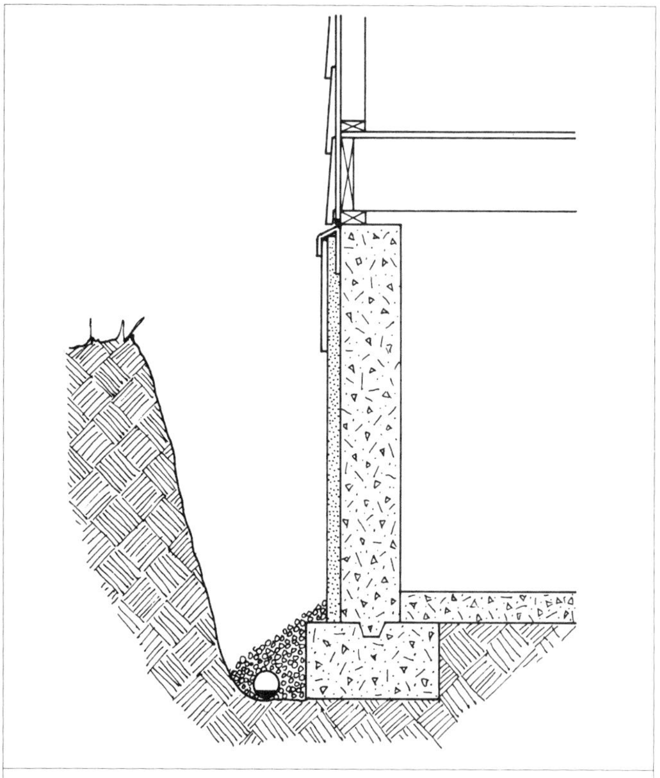

Exposing the foundation wall and drainage system.

Check the condition of the drainage tiles, repairing where necessary. Older homes are unlikely to have these installed. It is a good idea to install a system, but only if it can be done properly, draining to an appropriate discharge. It is best to do this after other work is completed. Consult someone experienced with drainage systems before proceeding.

Apply damp-proofing down from grade level to over the top of the footings. This can be done with two coats of damp-proofing compound.

Check all penetrations through the foundation wall. These will have to be sealed, removed, or extended out to accommodate the thickness of the insulation.

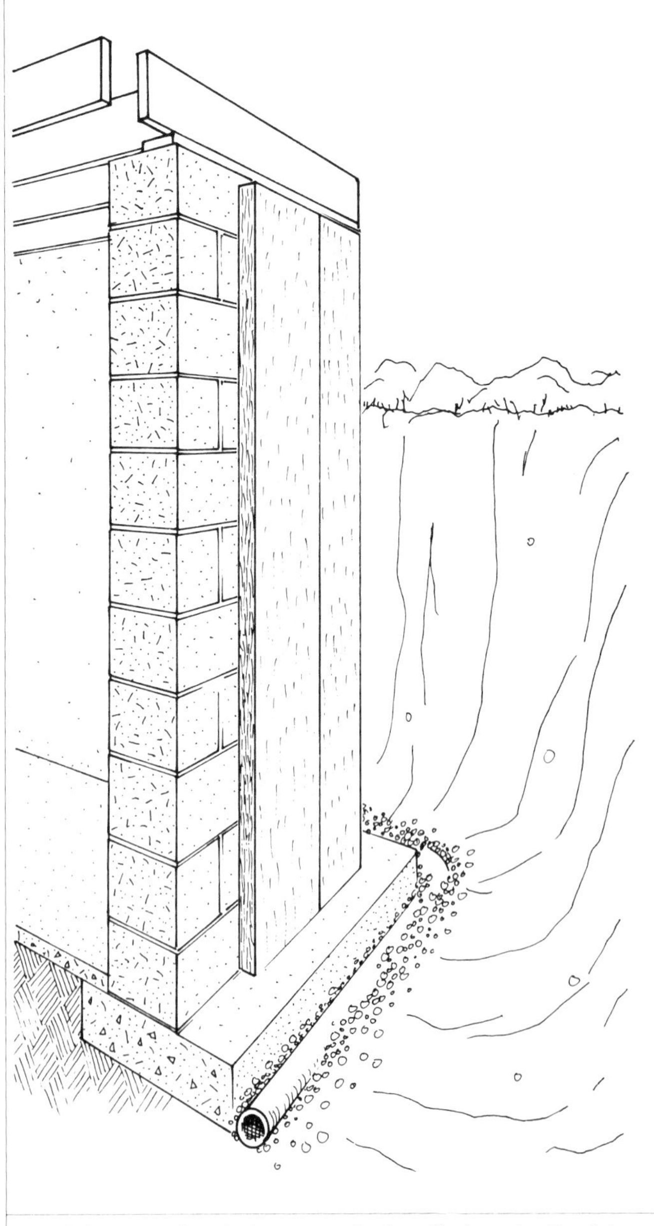

Draining-type insulations must be installed vertically right to the footings. A drain tile is essential.

3) Applying the Insulation

Three major types of insulation are used on the exterior of basement walls: rigid glass-fibre boards, polystyrene, or polyurethane/polyisocyanurate boards. See Chapter 2 for a description of these insulations. Rigid glass fibre and Type IV polystyrene boards are the materials most commonly used in exterior below-grade applications. Insulation which has drainage capability, such as glass fibre boards, must be used only if they are applied to the full depth of the foundation wall, if there is a drain-tile system, and if the insulation is applied with no horizontal seams.

Measure and cut the insulation to the desired height (generally from the top of the footings to the flashing). Start at one corner (overlapping at the corners) and keep the insulation sheets as tight to the wall as possible. Some people suggest using two layers of insulation with overlapping joints. The insulation is held in place at the top by the flashing and by corrosion-resistant fasteners and washers used to secure the protective finish. The below-grade portion of the insulation is held in place by the back filling.

It may be convenient (although more expensive) to purchase a special interlocking system of grooved polystyrene boards with steel channels. These should be used in the above-grade portion only to a depth of 300 mm (12 in.). There are also special clips and fasteners for applying the rigid board to the wall — check building supply stores.

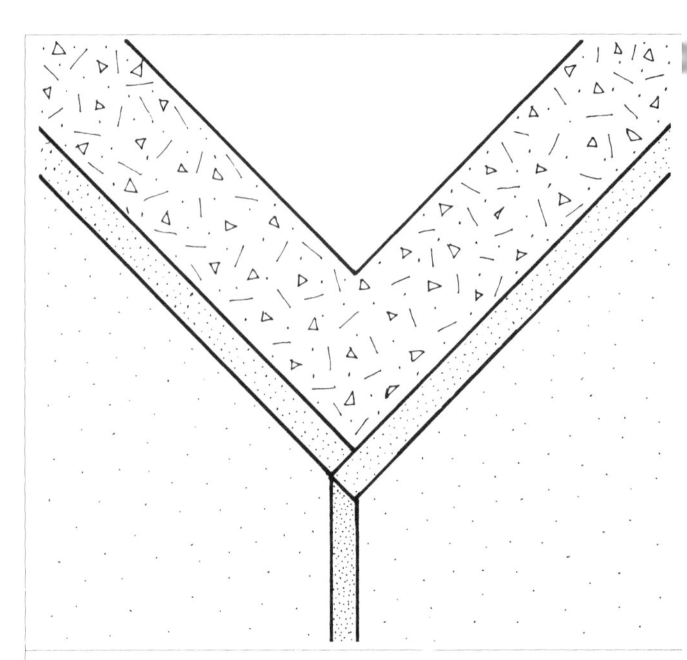

The insulation should overlap at the corners.

4) Flashing

Flashing helps keep the insulation in place, prevents water from getting behind the insulation, and provides a clean, neat junction. There are two major considerations: the location of the flashing which defines how far up the wall the insulation extends, and the type of flashing used.

If the siding can be partially removed or pried up, then standard "Z" flashing should be used. This is inserted at least 50 mm (2 in.) behind the siding and building paper.

If flashing cannot be inserted behind the siding (such as with brick) then either a metal "J" channel must be made up and installed prior to the insulation, or a wood flashing installed after the insulation. Flashing should accommodate the width of the insulation and protective coat. Wood flashing should be sloped with an overhang of at least 20 mm (3/4 in.), and have a drip edge on the underside.

Ideally, the insulation is carried up past the header area by at least 150 mm (6 in.). This often cannot be done because of practical or aesthetic reasons. If the insulation is only carried up to the header area or lower, then the header area should be air sealed and insulated from the interior. This is discussed later in this section.

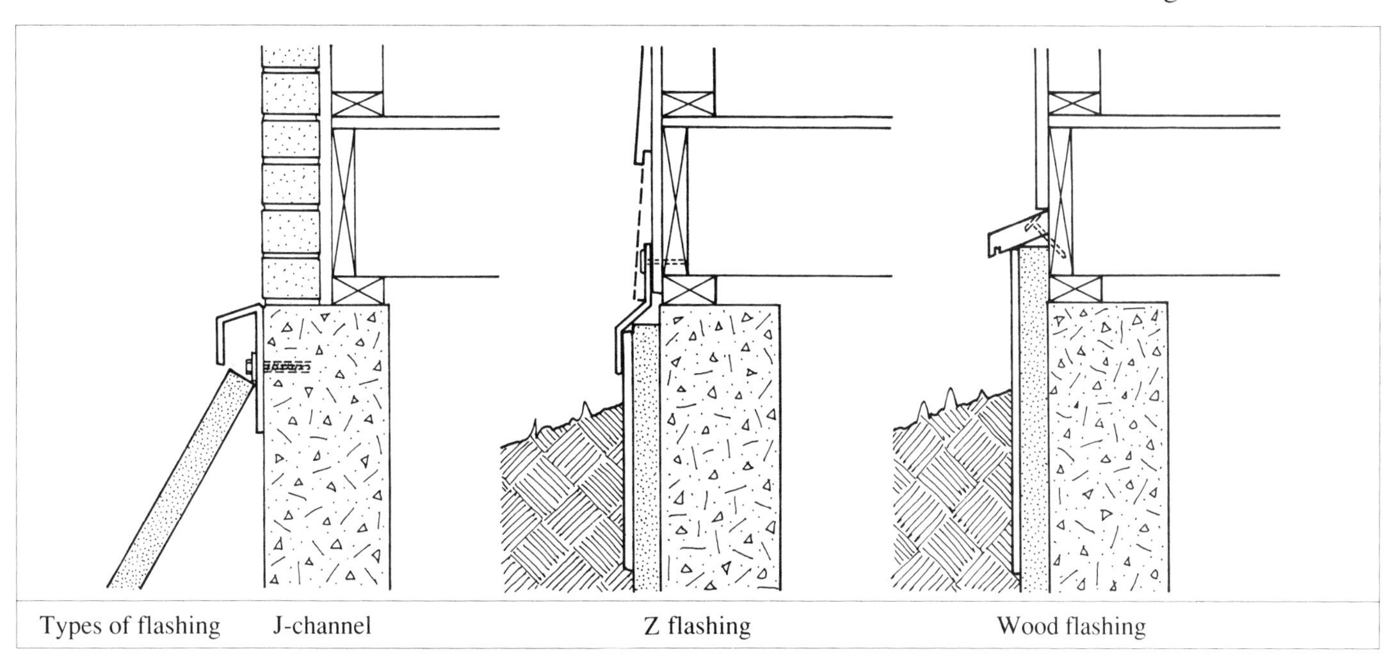

Types of flashing J-channel Z flashing Wood flashing

5) Exterior Protection

A covering is needed to protect the insulation from sunlight and damage by people and animals. It is applied from the top of the insulation to a point about 300 mm (12 in.) below ground level. Some possibilities include:

- Expanded metal lath with cement parging. (Follow the manufacturer's instructions.)
- Polymer-modified pargings. These go directly on some types of insulation without requiring metal lath. Check the manufacturer's literature.
- Pressure-treated plywood. This can be installed using stainless-steel fasteners 400 mm (16 in.) on centre.
- Fiberglass panels, or vinyl or aluminum siding which can match the house siding.

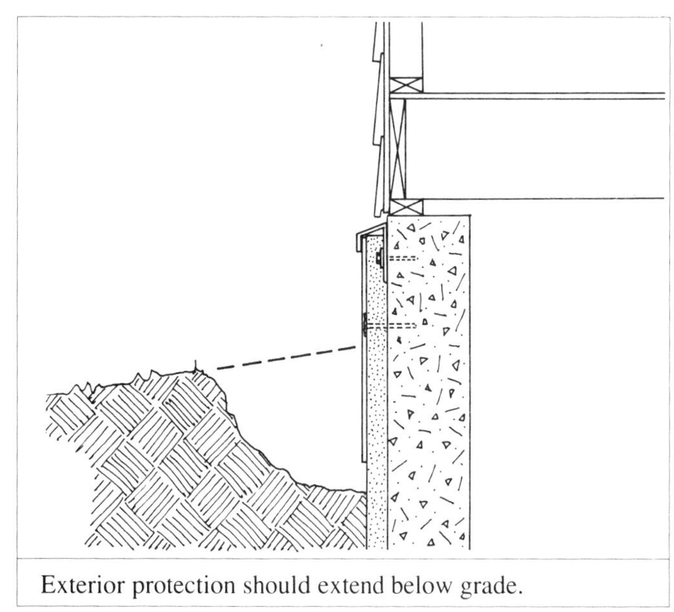

Exterior protection should extend below grade.

6) Backfilling

The drain tiles should first be covered with 150 mm (6 in.) of clean gravel (4 mm or larger) and preferably a strip of filter fabric. If a draining insulation is used, the gravel should extend at least 100 mm (4 in.) up the insulation.

Backfill in stages, removing large objects, and compacting or tamping the ground. If the soil is expansive clay or other poorly-draining type, it would be better to bring in free-draining soil.

When the hole is finally filled, make sure the ground slopes away from the house; usually a slope of 10 per cent — 200 mm (8 in.) for the first 2 m (6 ft.) is provided to allow for settling. This will encourage drainage away from the insulation, as will the addition of eavestroughing and downspouts. It is very important not to direct excess surface water towards the foundation.

The filled hole may be covered with any type of surface — patio stones, grass or a garden. Some additional settlement may take place — it is better to wait before undertaking any expensive treatments such as paving.

7) Finishing Details

In the case of wood flashing or the "J" channel, seal the joint between the flashing and the house with a suitable caulking. In the case of brick siding, the weep holes must remain clear.

Windows in the foundation can usually be finished by wrapping the insulation around the foundation to meet the window frame. Lath and parging can be applied over the insulation to the window frame. The joint between the frame and parging is caulked. This will have to be inspected periodically.

Doors should be outlined with a "J" channel or equivalent flashing. The door sill may have to be extended to protect the flashing beneath the door.

Seal penetrations through the insulation and covering to prevent wind and water entry. Some penetrations (gas regulators, electrical conduit) may be better left uninsulated.

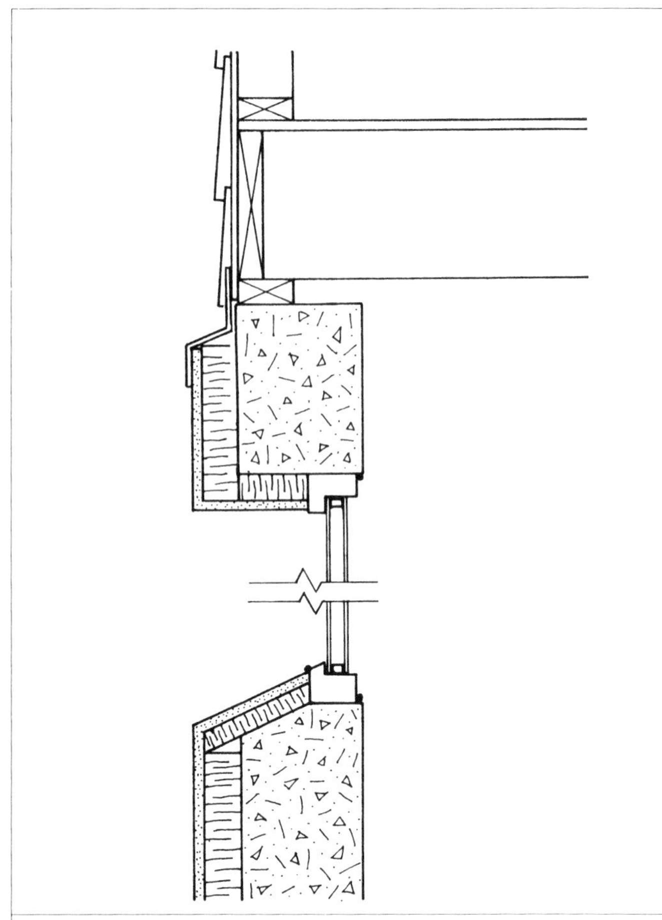

The sill should slope away from the window.

8) Insulating the Header Area

If the exterior insulation does not extend above the header area by at least 150 mm (6 in.), the header area should be air sealed and insulated from inside the basement. Caulk the area where the wood frame wall meets the concrete foundation wall at the sill plate with a good caulking compound such as butyl rubber or polysulfide to provide the air barrier. The header can be insulated with batt and rigid board insulation as described in Part II.

Complications

Part of the Basement Wall Encloses a Cold Cellar or an Unheated Garage.

Apply the insulation inside the basement, against the cold cellar or garage walls. The wall should be treated as if it were an exterior basement wall. The doorway from the basement should be weatherstripped and insulated. Finally, insulate the ceiling of the cold cellar or garage.

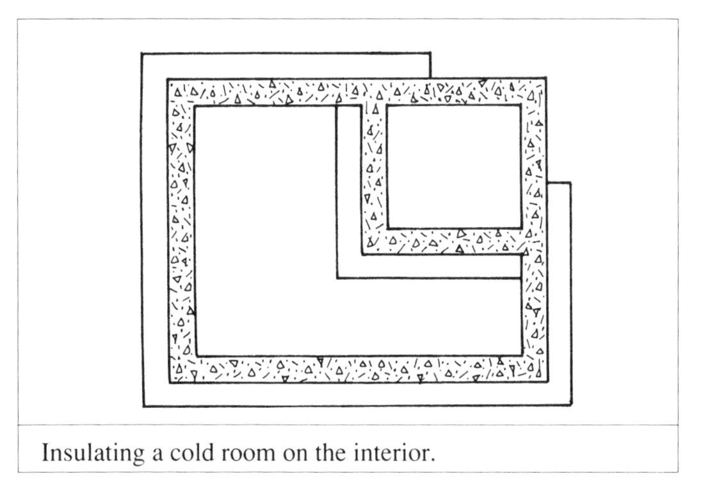

Insulating a cold room on the interior.

The outside insulation around the rest of the basement should be extended at least 0.6 m (2 ft.) beyond the inside wall junction (as illustrated) to minimize heat loss at these points. The further the insulation overlaps, the better.

There is a Concrete Porch Butted against the Basement Wall, a Paved Driveway, or Some Other Obstruction.

Once again, the insulation should switch to the inside around these obstacles. There should be at least 0.6 m (2 ft.) of overlap, to provide continuous coverage and reduce the heat loss through the thermal bridge.

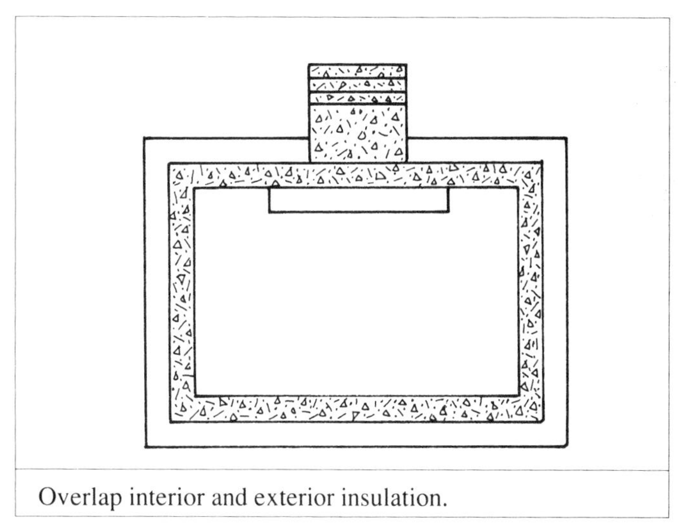

Overlap interior and exterior insulation.

Part II How to Insulate inside the Basement

Assessing the Situation

Whatever the basement type, there are a number of things to examine before beginning work since the type and condition of the wall will influence how you insulate.

- Interior features that may make insulating more difficult (uneven walls, stairs, services, cupboards, partition walls, etc.).
- Indications of structural problems (cracks, bulges).
- Indications of moisture problems (leaks, dampness, efflorescence, blistering paint).
- Insulation requirements: type, level, location.
- Finishing details required.

Before adopting a system for applying insulation and a protective finish to your basement walls, you may want to consult a local builder and check with local building authorities to be sure that your project will meet the requirements of building codes. Also, some regions have particular problems, such as frost heave due to expansive clay soils, and these special problems should be considered before beginning work.

The choice is basically between two types of insulation — batt/blanket insulation, and rigid plastic or glass fibre insulation. Refer to Chapter 2 for detailed information on these types of materials.

The rigid plastic board insulations generally have a higher RSI value per millimetre than the batts and therefore require less basement space and a thinner supporting framework. They are less prone to moisture damage than the batts. However, they are more expensive and must have a fire-resistant covering.

In some areas of the country you can purchase special basement insulation systems that may allow for a more convenient or efficient use of materials. Check with your building supply dealer. Remember, when choosing an insulation system it is important to calculate the cost of the whole system under consideration — including the price of support materials (studs and fasteners), the air and vapor barrier, a fire-protective covering and installation costs.

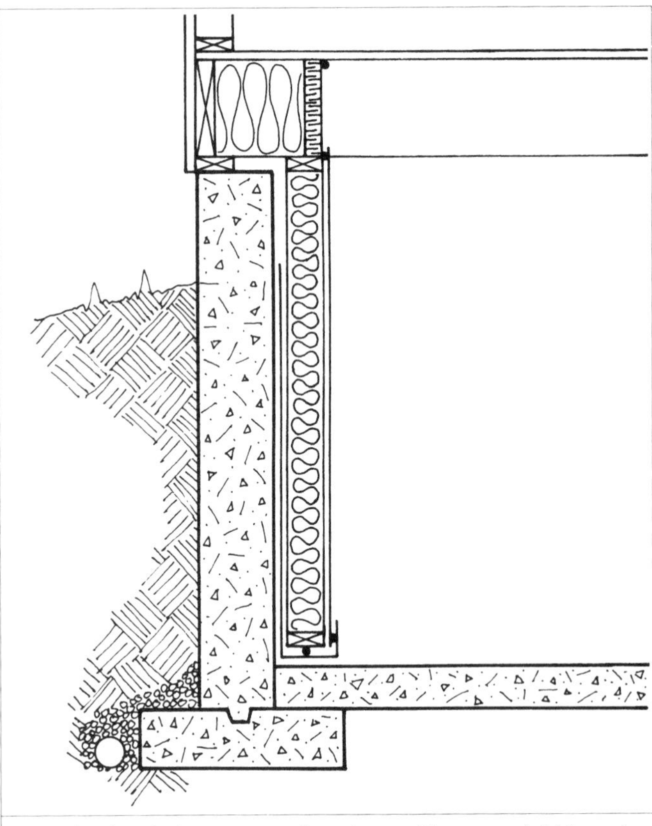

Interior insulation can use framing and batts or rigid board insulation with an attachment system.

Special Safety Considerations

Refer to Chapter 2, Part IV, for general advice on safe working procedures. When working inside basements remember to follow these guidelines.

- Provide adequate temporary lighting.
- Keep yourself and the insulation materials away from the flue pipe of the furnace and any other sources of heat.
- Watch out for older wiring, such as the knob-and-tube type, that may be in poor condition. This is a common hazard when working in older basements.

How to Insulate Inside the Basement Using Rigid Board Insulation

This method works best if the basement wall is even and vertical, as the board material is fairly rigid. It is usually restricted to concrete block or poured concrete walls. Rigid insulation panels are secured to the concrete using mechanically-attached nailing strips, and the assembly is protected with 12.7 mm (1/2 in.) gypsum drywall secured to the nailing strips.

Preparation
After you have checked the wall and made any necessary repairs, air-seal all leakage paths, such as at the sill plate, and around penetrations. This step is very important as this provides the primary air barrier system.

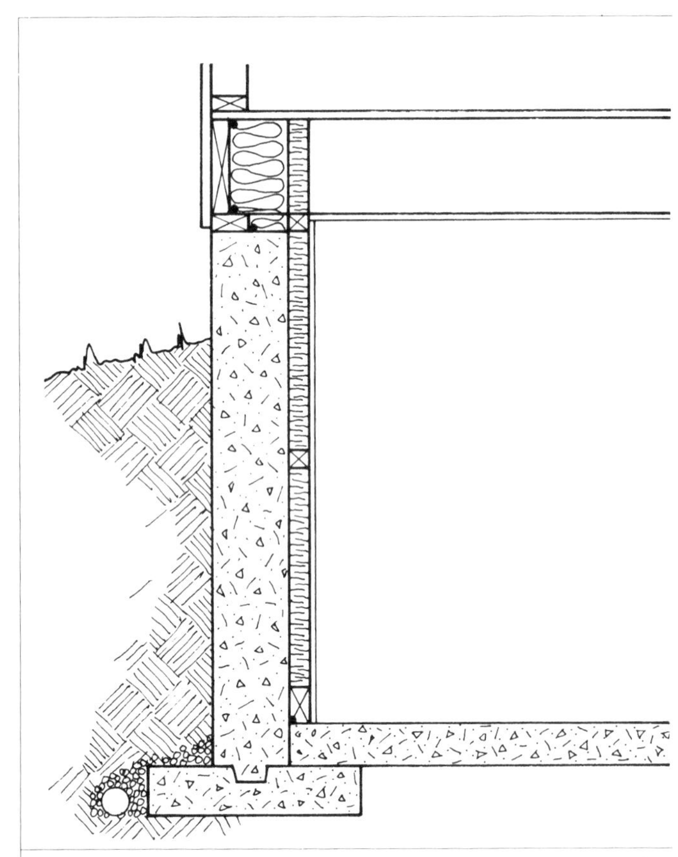

Rigid board insulation involves: 1) air sealing the old walls; 2) installing the insulation; 3) finishing.

Installation
Mechanical fasteners are essential to secure the gypsum board to the wall. This often means that wood nailer strips are used, with the insulation placed between or behind the strips. Alternatively, the insulation can be held in place with a special wood or metal nailing strip that fits within grooves or notches cut in the insulation panels. In either case, the drywall is screwed or nailed in place into the nailing strip. Corrosion-resistant concrete fasteners secure the nailing strip into the concrete wall. The insulation can be glued on the wall as a temporary measure until the gypsum board is mechanically fastened to the wall or nailing strips.

The best system for you will in part depend on the the type of finish you will install and the loading requirements on the finished wall. For example, vertical panelling will have different requirements from horizontally installed drywall. Ask at your building supply outlet for the different options available.

You should install at least RSI 2.1 (R12). Consider installing it in overlapping layers to minimize heat loss through the wood nailing strips. Be sure to install the insulation snugly, to eliminate air circulation at the edges.

Finishing

The joist area must also be insulated and sealed (unless the joists are embedded in concrete). See the next section for details.

The entire wall must be protected with 12.7 mm (1/2 in.) gypsum board or equivalent fire protection. This includes the joist space if a new ceiling is not installed.

If you purchase a prefabricated system with an attached fire-protective interior finish, you will have to carefully follow the manufacturer's instructions.

How to Frame and Insulate Inside the Basement — Step by Step

This system consists of a new wood frame wall with batt or blanket insulation, an air and vapor barrier, and finishing. It can provide high levels of insulation at a relatively low cost.

Preparation

Before you begin, be sure to caulk any cracks between the foundation and the sill, as well as any other air leakage paths. See Chapters 2 and 3 for a description of the best materials and techniques.

Inspect the walls for possible moisture problems. Occasional dampness on the basement walls (especially in late spring or early summer) is permissible as long as the correct procedures are followed when installing insulation.

When water leaks are major or frequent, the source of the problem must be corrected and the wall repaired. It may be preferable to excavate, dampproof, and insulate from the outside.

Cover the basement walls with a polyethylene **moisture barrier** that extends from grade level only to the bottom of the wall allowing extra at the bottom — about 300 mm (1 ft.) — to lie under the new frame wall. This will protect the insulation, studding and wall finish from possible water damage.

Framing a New Wall

The next step is to install a wood frame wall in the basement. There are two approaches. You can install the new wall flush to the old wall using 38 mm x 89 mm (2x4) lumber. Alternatively, you can use 38 mm x 64 mm or 38 mm x 89 mm (2x3 or 2x4) lumber built out from the wall by 64 mm (2 1/2 in.). The second method takes up more room, but does provide more insulation, less thermal bridging through the studs, and better moisture protection.

Because the wall frame is built out from the wall, the studs will not touch the cold exterior walls and there will be space for an extra layer of horizontal insulation.

The bottom plate should sit directly on the extension of the polyethylene moisture barrier. If you never have moisture leakage on the wall to be insulated, then you can also set the bottom plate on a bead of caulking compound or gasketing material to create a tight seal.

Next, fasten the top plate to the bottom of the joists. Where the wall runs parallel to the joists you will have to build in a nailing support for the top plate (the approach you use will depend upon your par-

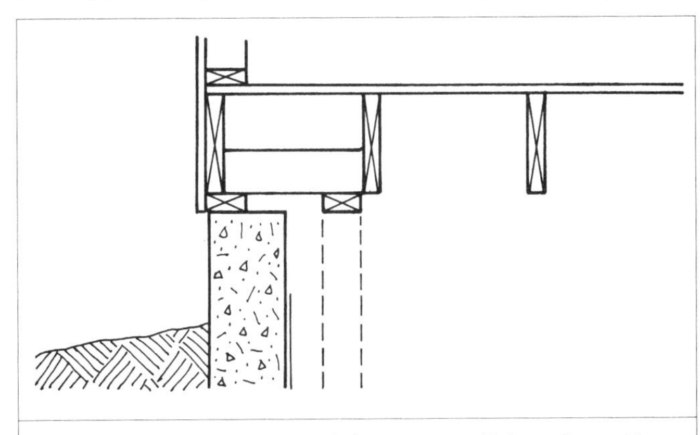

Top plate detail where the joists run parallel to the wall.

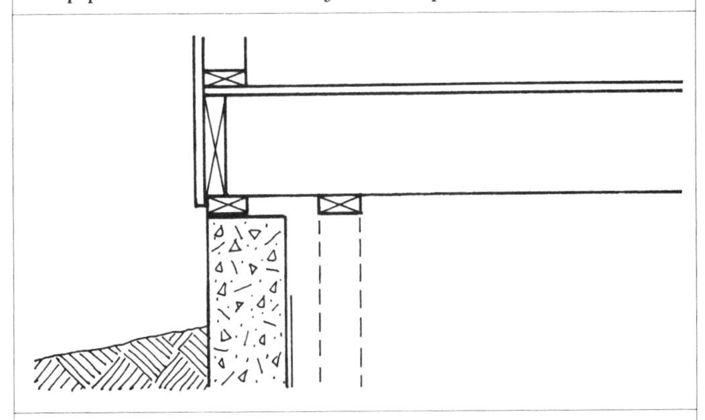

Top plate detail where the joists run perpendicular to the wall.

ticular house). Now is the time to line up the new wall properly using a plumb bob or level and straight edge.

Then install the studs 600 mm (24 in.) on centre (i.e. from the centre of one stud to the centre of the next) for 38 mm x 89 mm (2x4), or 400 mm (16 in.) on centre for 38 mm x 64 mm (2x3) lumber. The choice of spacing will depend upon how much structural support your finishing material needs. Make sure that the studs are perfectly vertical and accurately spaced so that the insulation will fit in snugly and the finish can be installed without problems. Measure each stud separately! Extra framing is needed around windows and doors in the foundation.

If you are a good framer and all alignments are perfectly level and square, you may be able to build the wall on the floor, tilt it into place, shim the bottom plate, and then secure it.

Use dry lumber for the framing. If not, allow the framing to dry for at least two weeks **before** adding insulation and covering the wall with the air and vapor barrier. Some temporary bracing may be tacked on to keep the wet studs from twisting as they dry.

Insulating

Insulation is installed in two layers. The first is a horizontal layer between the studs and the wall. **It is very important that the insulation is tight against the foundation wall**. Next, install a layer snugly, vertically between the studs. There should be no gaps or air spaces where convection currents could help heat to bypass the insulation. Insulation should be wide enough to match the stud spacing, and be cut to fill the full height of the wall.

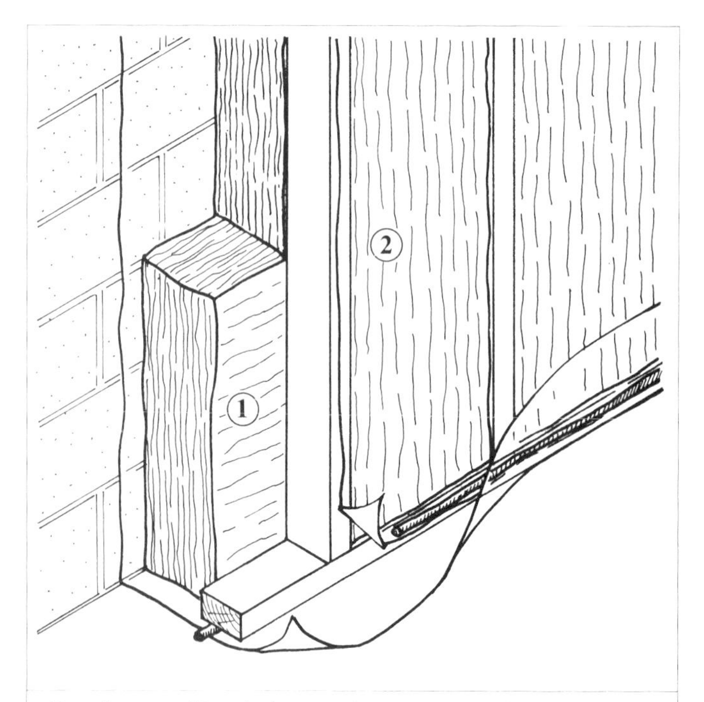

Two layers of insulation can be used: 1) horizontal between the foundation wall and the studs; 2) vertical between the studs.

Finishing

Install a polyethylene air and vapor barrier over the studs and insulation. Leave enough extra at the top to connect to the air barrier in the joist space as discussed below. Seal all edges, seams and penetrations in the barrier with acoustical sealant. All joints should overlap over a stud and be sealed with a continuous bead of sealant that is run between the layers of polyethylene at the lapped joints. Staple the polyethylene to the stud through the bead of sealant.

Except for foundations where the joists are embedded in concrete, the joist space should be insulated and sealed. This can be done by filling the space with a section of glass fibre insulation and then installing a piece of low-permeability rigid board insulation, cut to size, in each joist area above the top plate of the new frame wall. Seal the edges of the rigid insulation with non-hardening sealant in order to get a tight seal. The polyethylene air and vapor barrier from the wall is sealed to the bottom of the rigid insulation.

The same basic technique can be used where the joists run parallel to the wall. The last joist that runs above the sill is usually inset from the foundation wall. Depending on the layout of the house, this space can be fitted with long strips of rigid insulation. The polyethylene air and vapor barrier installed on the wall is then lapped up and sealed to the rigid insulation.

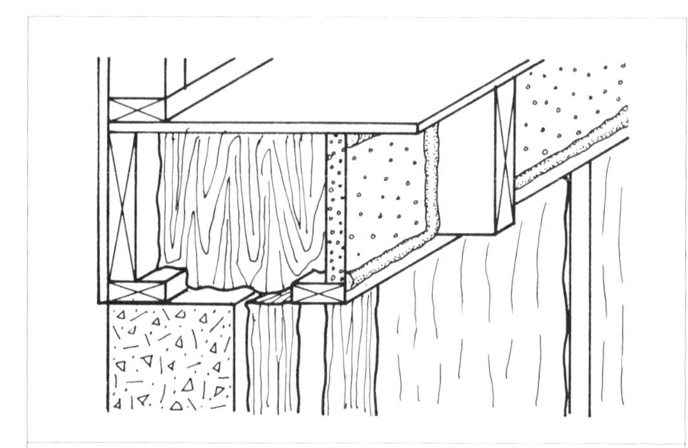

Rigid extruded polystyrene insulation caulked between the joists continues the air and vapor barrier up to the subfloor.

Finally, nail up the finishing surface. Make sure it is **tight** against the insulation. The polyethylene barrier and rigid insulation will represent a fire hazard until covered. This also applies to the joist spaces.

If the joists are embedded in concrete, it is better not to insulate the joist area. This will help to keep the joist ends warm and dry. However, you should still air seal around each joist.

Complications

Wall Space Interrupted by Pipes, Ducts, or an Electrical Panel.

- Move water supply pipes away from the wall if at all possible. If they cannot be moved, install the insulation and the air and vapor barrier behind the pipes so that they are on the warm side. Never place insulation in front of the pipes. Any pipes that pass through the air and vapor barrier should pass through a plywood board that is sealed to the main air and vapor barrier and the gaps around the pipes caulked.

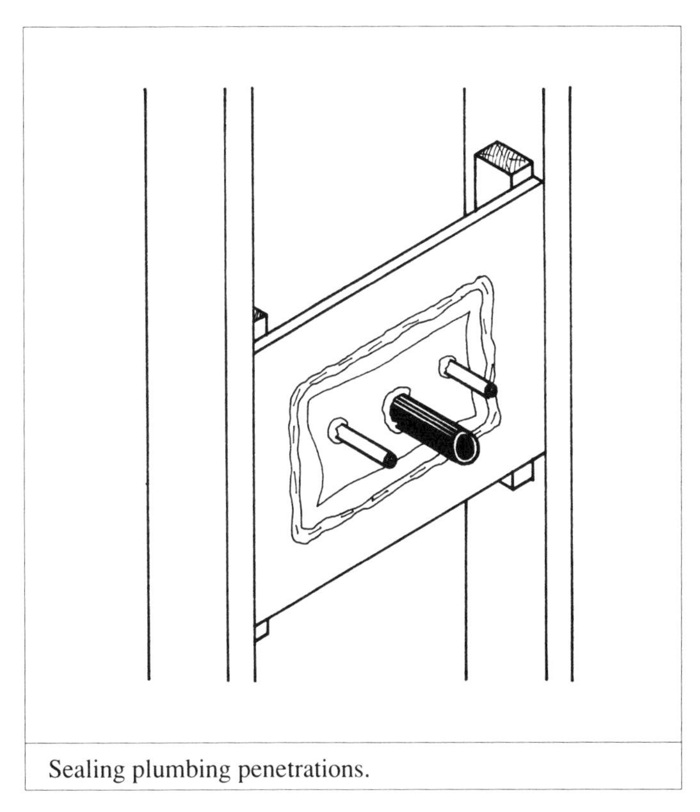

Sealing plumbing penetrations.

- Do not insulate around any flue pipes. Different clearances are required depending on the type of flue. Check with the manufacturer or a heating system specialist. Similarly, furnaces, wood stoves and fireplaces require clearances from the wall. Do not insulate if you cannot maintain the proper clearance.
- Be careful working around the main electrical panel. Even when the power to the rest of the house is off, **the panel will still be "live"**. It's better to have an electrician move the panel out to accommodate the new wall.

Basement Wall Interrupted by a Window.

- Seal the point where the window frame adjoins the wall with caulking compound, and then insulate.

Basement Wall is Very Irregular.

- An irregular basement is usually made of stone or rubble and is rarely damp-proofed on the outside. An interior retrofit is not recommended, but where no water or moisture problems exist, it may be possible to insulate on the inside. Batt insulation will conform to the irregularities. When insulating on the inside of stone and brick foundations, do not insulate the joist space. This will allow moisture in the foundation wall an escape route. The joist area should still be air sealed.

Part of the Basement is a Cold Cellar or an Unheated Garage.

- Apply the insulation to the cold cellar or garage wall separating the heated basement from the unheated space, as if it were an external basement wall. The doorway from the basement should be weatherstripped and insulated. Finally, insulate the ceiling as described in Part III for unheated crawl spaces.

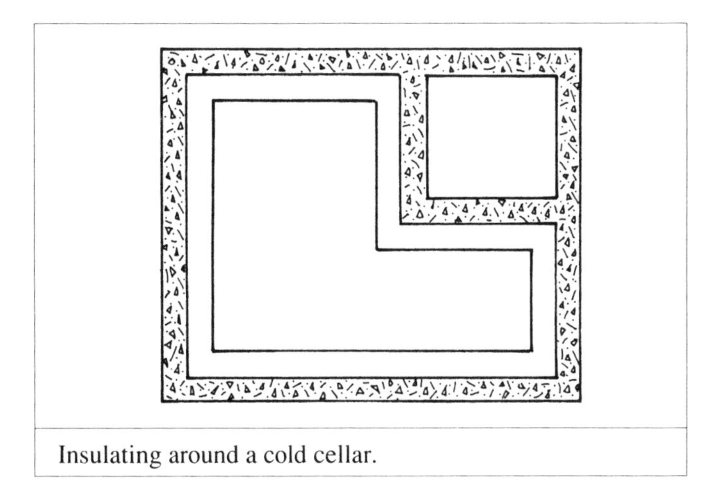

Insulating around a cold cellar.

If you insulate your cold cellar, there will be an added benefit — it will be colder! Keep a check on the winter temperature so that adjustments can be made to prevent freezing.

The Basement Has a Pony Wall

A pony wall consists of short sections of wood frame wall sitting on top of a conventional concrete foundation. In this case, the wood frame section is insulated between the studs, and the concrete section is insulated on the interior (assuming there are no moisture problems). The insulation on the concrete is extended up about 200 mm (8 in.) to overlap with the frame section. A ledge is created at this point (see illustration).

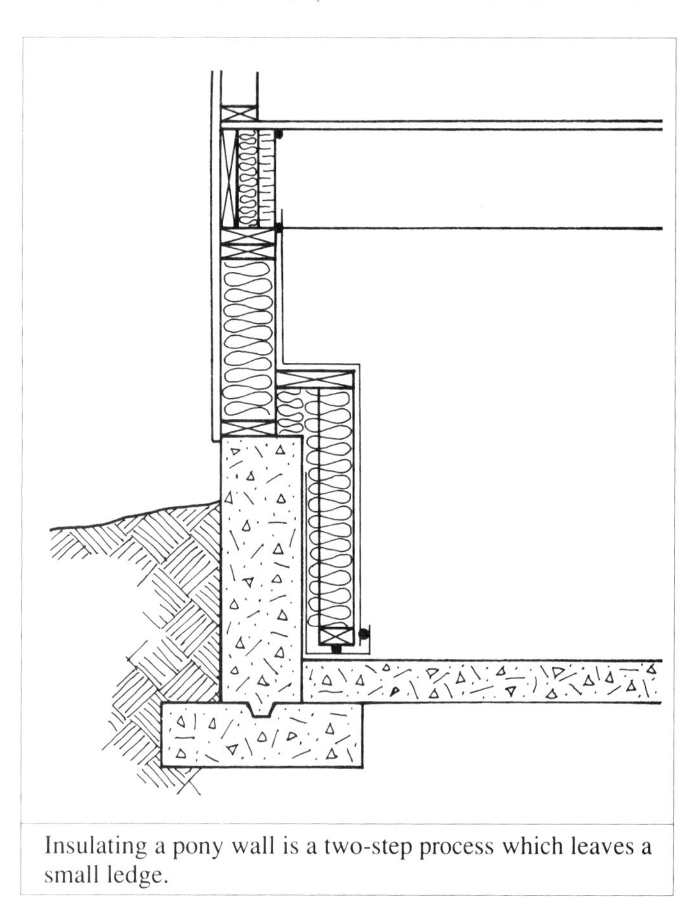

Insulating a pony wall is a two-step process which leaves a small ledge.

Overhangs

An overhang over the foundation should be air sealed and insulated. It is usually possible to remove the finish underneath the overhang and air seal the space between the joists above the foundation with polyurethane foam or caulked, low-permeability rigid insulation. Insulate the joist space with batt or blanket insulation before reinstalling the finish. Heavyweight building paper, olefin sheet, or other breathing-type air barrier can be installed before the finish.

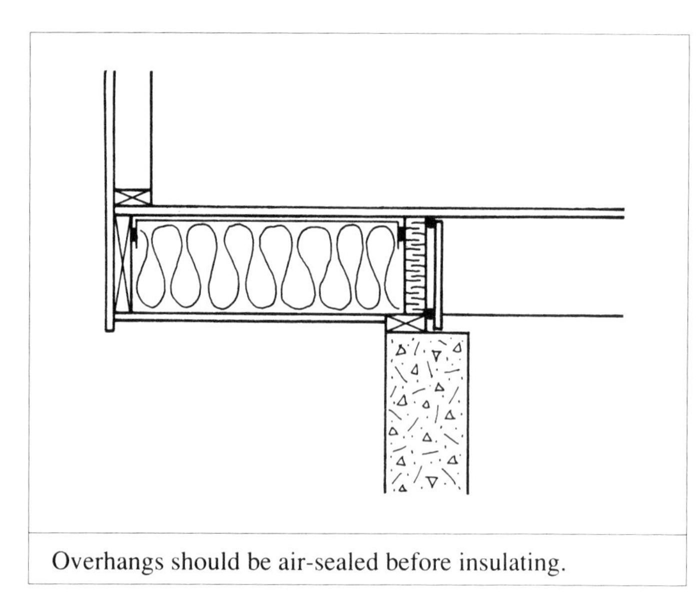

Overhangs should be air-sealed before insulating.

Part III Crawl Spaces

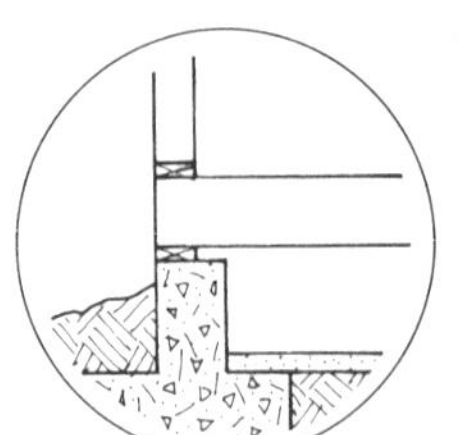

Theoretically, a crawl space could be insulated in either of two ways:

- The walls of the crawl space could be insulated, on either the inside or the outside, creating a heated area.
- The house floor above it could be insulated to keep heat from getting into the crawl space in the first place.

Insulating the walls is recommended for the following reasons:

- Heating ducts and water pipes in the crawl space won't have to be insulated, and won't freeze. Any heat loss from the ducts will not be a total waste.
- The walls can be insulated on the **outside** to reduce the internal moisture problems which can develop in damp crawl spaces, and keep the soil below the footings warm.
- It is usually **easier** to do a better insulating job on the walls, especially when the crawl space is shallow or the joist spaces are uneven or oddly shaped.
- **Less material** is usually required if the crawl space is of a typical height — less than 1.5 m (5 ft.).

Moisture Barriers

If there is no moisture barrier on the crawl space floor, add one. The barrier should be 0.10 mm or 0.15 mm (4 mil or 6 mil) polyethylene overlapped at the seams and held down with occasional old boards or some other scrap material. If there is likely to be any traffic in the crawl space, you will need to protect the polyethylene with a 50 mm (2 in.) layer of sand.

How To Do It: Crawl Space to Be Insulated and Heated

From the Outside

- Insulate the outside wall exactly as described for the outside basement wall in Part I.

- If outside obstructions (a porch, a paved driveway, etc) make it impossible to completely encircle the crawl space from outside the house, then the inside of the wall may be insulated at those points.

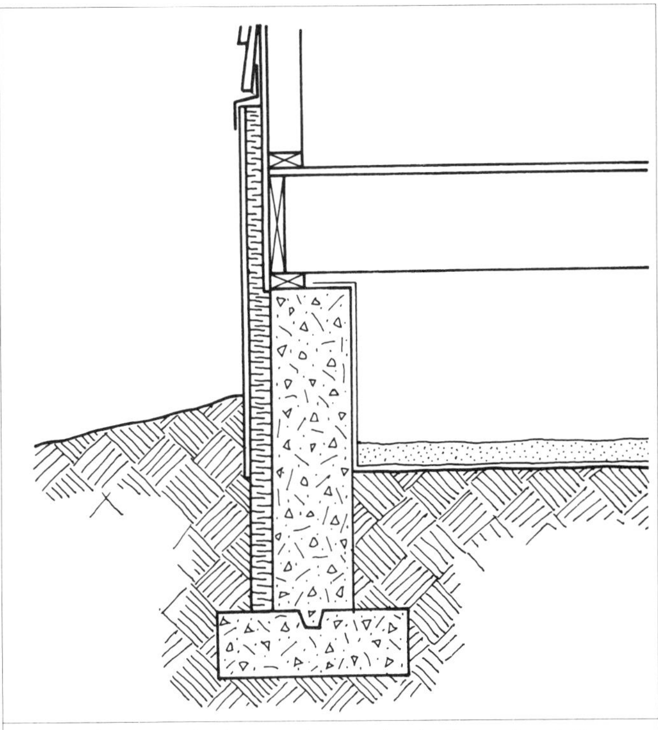

Insulating outside the crawl space is similar to insulating a full basement.

Make sure the inside and the outside portions overlap by at least 600 mm (2 ft.), as illustrated, and insulate the inside portion following the instructions as described below.

- If your crawl space doesn't open into a full basement, it should have ventilation at a ratio of 1 to 500 (vent area to floor area). The best time to ventilate is in the spring since summer ventilation can increase condensation. Make sure these vents are closed and well sealed and insulated each winter!

From the Inside

- If using polystyrene or semi-rigid glass fibre insulation, insulate in the same manner as outlined for the inside of a basement, in Part II.

- Apply a polyethylene moisture barrier to the crawl space floor and install adequate ventilation as described above.

Note: If the foundation footings are above the frost line, insulate on the *outside* of the crawl space walls. By insulating on the outside, the walls will be kept warmer, avoiding any possibility of frost heave. Shallow footings can be kept warmer by placing a layer of horizontal insulation sloping away from the foundation.

Make an effort to keep water **away** from the foundation walls (slope the ground away from the house and install eavestroughing where necessary).

How To Do It: Partially Heated Crawl Space

It is possible to insulate between the joists and create an unheated crawl space. However, this can lead to problems of freezing pipes, frozen ground, and possible rot at the joist ends. For these reasons, floor insulation is only recommended when combined with foundation wall insulation to create a partially heated crawl space.

A few points of general importance should be summarized:

- The air and vapor barrier must be applied on the **warm** (top) side of the insulation. If the floor above the crawl space is already covered with an impermeable material (e.g. linoleum or plywood) you already have a vapor barrier where you want it. The solid materials of the floor can serve as the air barrier, but be sure to locate and seal any air leaks. The air tightness at the perimeter joist spaces is critical. This area can be sealed with polyurethane foam.

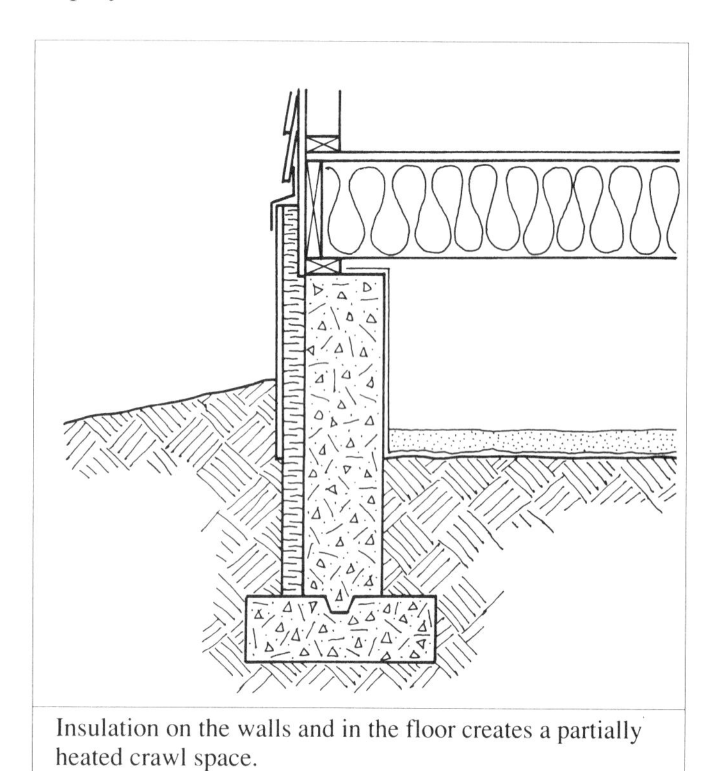
Insulation on the walls and in the floor creates a partially heated crawl space.

- Batt insulation may be held in place with heavy-duty permeable building paper stapled to the joists, or by chicken wire, sheets of polystyrene beadboard, or a commercially available insulation support system.

- Place the insulation firmly against the floor above. It should be installed so that it fills the space between the subfloor and the support system (usually the depth of the joists).

- Tape the seams in any heating ducts and **insulate** all ducts and water pipes in the crawl space. Remember, even insulated water pipes may freeze if the temperature of the crawl space is allowed to fall below freezing.

- Make sure the crawl space is adequately ventilated in the spring. Vents should be installed at a ratio of **1 to 500** (vent area to floor area). Do not ventilate in winter: the vents should be plugged and insulated.

- There must be a moisture barrier on the crawl space floor.

- If your basement has both a full basement and a section of crawl space where the floor has been insulated, remember to insulate the wall separating the basement from the crawl space.

Note: If the ground level inside the crawl space is lower than the ground level outside, there is a slight danger that frost heave can damage the walls by pushing them inward. Make every effort to keep water **away** from the foundation walls — slope the ground away from the house and install eavestroughing where necessary. As an added safety precaution if freezing becomes a problem, you may want to install a thermostat attached to a small heater in the crawl space. This unit can turn on and heat the walls when the crawl space temperature approaches freezing.

Part IV Open Foundations

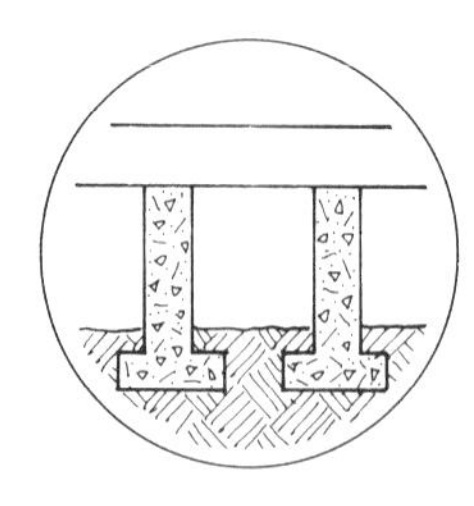

Some older homes and cottages have open foundations. They should be insulated between the joists in the same way as crawl spaces. There should be a good air seal and the insulation should be protected from the wind and animals. It may be possible to build an insulating skirt around the foundation, creating a heated crawl space.

If the joist space is already covered, the insulation may have to be blown in. Read the comments in Chapter 4, Part II that deal with insulating the attic floor. The open foundation situation is virtually identical, with the following exceptions.

- The vapor barrier is placed above the insulation instead of below.
- It is crucial that the insulation is blown in at a high density so that there is no air space between the floor above and the insulation.

Part V Concrete Slab on the Ground

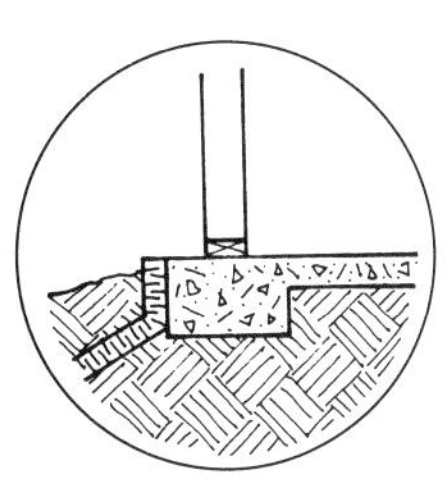

Insulation is applied to the foundation of these basementless houses exactly as you would insulate the outside of a full basement. Refer to Part I of this chapter for complete details.

If the foundation is shallow on frost-susceptible soils, a near-horizontal layer of insulation should be installed. The closer the footing is to the surface, the longer the horizontal insulation should be. Consult your local buildings department or the insulation manufacturer for detailed information.

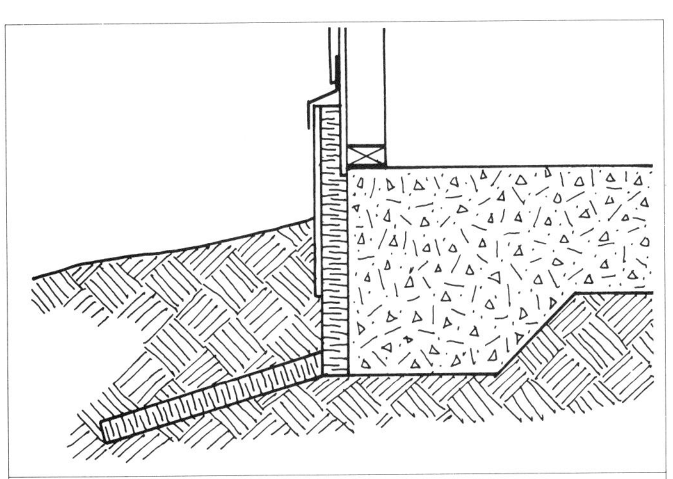

Horizontal insulation provides frost protection to the slab on grade.

Contracting the Work

Exterior Insulation

Contracting for the excavation can save time and a backache. If the whole job is being contracted out, ensure that the quote includes parging, damp-proofing, drain tiles and aggregate where required, the type and quantity of insulation used, the fastening techniques used, the sealing of penetrations, the flashing and finishing details, a sloping grade, and clean-up.

Interior Insulation

Quotes for interior insulation should include details on wall preparation, the installation of a moisture barrier, the type and spacing of the framing, the type and quantity of insulation, details of a sealed air and vapor barrier, insulating and sealing details at the header, sealing of penetrations, and finishing materials.

6 Insulating Walls

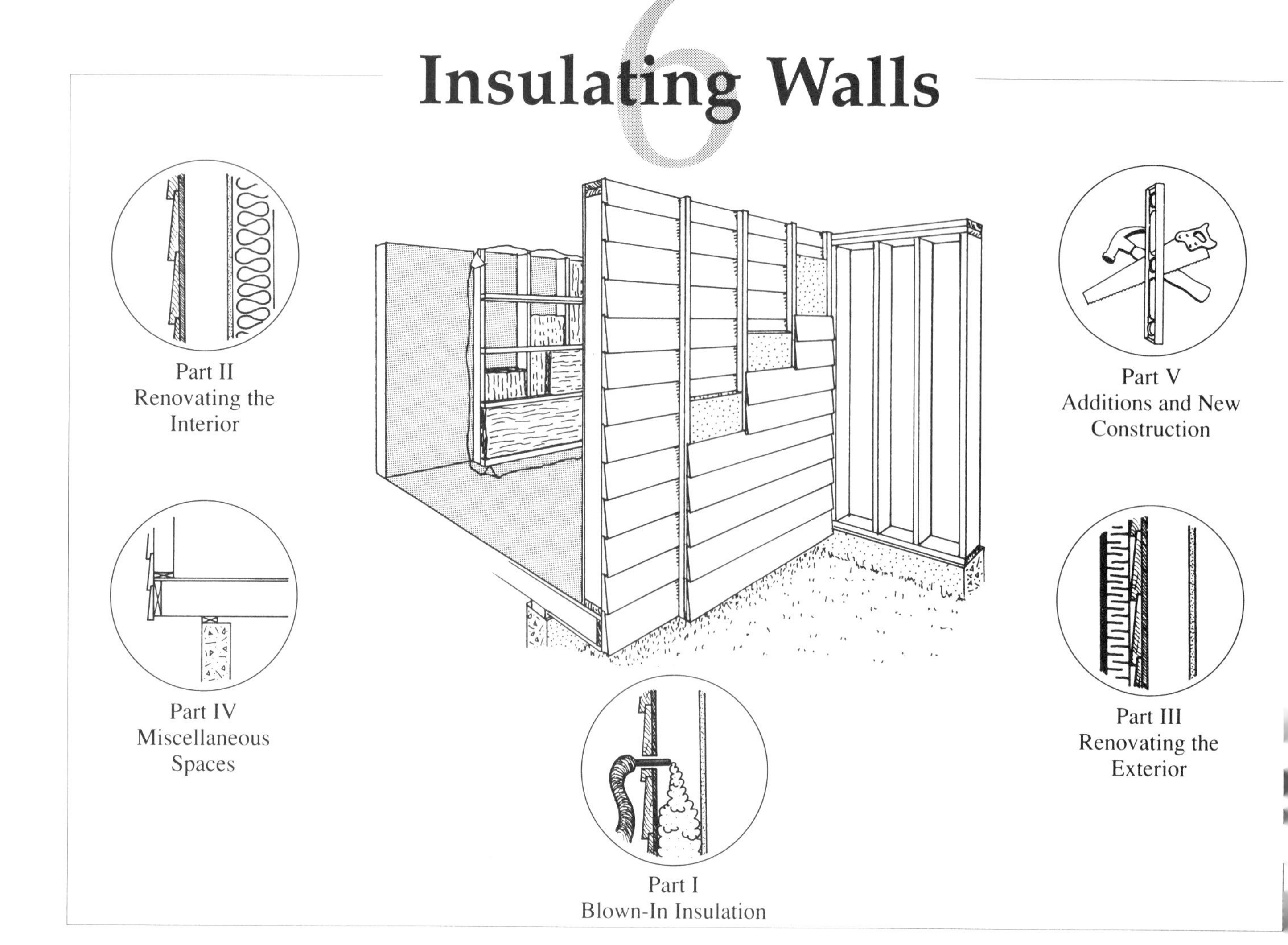

Introduction

Walls can account for 10 to 30 per cent of house heat loss. In addition to heat loss through the wall, there are many cracks and penetrations which allow uncontrolled air leakage into and out of the house.

Types of Wall Construction

Solid Walls
Solid walls include brick, concrete block, log and wood plank. Solid walls do not have a cavity that can be insulated. The only way to insulate these types of walls is to add insulation to the exterior or to the interior. Many solid walls do have a small cavity, generally under 25 mm (1 in.), that is used to collect and drain water out of the wall. Never insulate these cavities or plug their drain holes.

Concrete block walls usually have hollow cores which allow air circulation. The cores cannot effectively be insulated, since the block and mortar will continue to act as a "thermal bridge". However, air can easily circulate inside the block cores, increasing convective heat losses. Seal all possible air leakage routes into the blocks.

Frame Walls
Frame walls have a cavity that may be insulated. Different construction techniques determine the size of the cavity and ease of access from either the interior or exterior. The wall construction also affects details that can interfere with the insulation, including top and bottom plates, fire stops, blocking, plumbing, wiring, and heating ducts.

> **Note:** Frame house with **brick veneers** usually have a 25 to 50 mm (1 to 2 in.) air space between the bricks and the frame wall for drainage. While the larger cavity in the frame wall can be insulated, the drainage cavity behind the brick veneer should never be insulated.

Opportunities for Upgrading

Empty cavity frame walls are the easiest to insulate. Insulation can be blown in from the top and bottom, or from the interior or exterior.

Frame walls with some insulation or solid walls are more difficult to insulate, although they should be air sealed as described in Chapter 3. They can be insulated as part of a major repair job or renovation. From the inside this includes wall repairs, drywall replacement and decorating and any renovation changes to the wall. From the exterior, insulating can be combined with re-siding.

Solve any moisture or structural problems before insulating. Indications of problems include staining, dripping, mold growth, cracks on the inside wall finish and in the exterior siding, and windows and doors that don't operate properly because they are out of square.

Note: It is important to consider both vapor barriers and air barriers, especially when extending an existing wall from either the interior or exterior. **Keep in mind that in most climate zones, the vapor barrier must be on the warm third of the finished wall.** See Chapter 1, Parts III and IV for more details. You must also consider the location and condition of old vapor barriers, which could be as simple as plaster walls with several coats of paint.

Part I Blown-In Insulation

If you have empty wood-frame wall cavities, have a professional insulation contractor blow in loose-fill insulation. Before you proceed, however, remember that the stud space is likely only about 90 mm (3 1/2 in.) thick. If there is already 50 mm (2 in.) of insulation, the benefits of blowing in more will be small and it will be very difficult for a contractor to do a good job. You may have to do a little exploring at different locations to find out what is in the walls. Try looking behind electrical outlets (with the power off!).

A small hole must be drilled into each stud space in the walls; in most cases, two or more must be drilled per storey — not more than 1.5 m (5 ft.) apart vertically, and above and below windows and doors.

There are three possible ways of doing this:

- **From the inside.** Small holes of 15 mm to 50 mm (5/8 in. to 2 in.) are drilled in through the **inside** wall finish and the insulation is blown directly into the wall.

  The holes must be completely sealed after the job is done. The patch job can be messy and unsightly unless immediately covered with new paint or wallpaper, that should be impermeable.

This approach works best when combined with redecoration or renovation. In fact, if the interior finish is in bad condition and needs to be replaced or recovered, it should be possible to drill the holes, blow in the insulation, plug the holes, install a well-sealed polyethylene **air and vapor barrier** over the old interior wall, and apply new drywall over that.

Blown-in insulation from the exterior.

- **From the outside**. Most types of exterior siding can be drilled or lifted to permit access to the stud wall behind. Brick siding can have single bricks temporarily removed. This leaves sufficient space to repair holes in the sheathing. Ideally, two stud spaces can be filled from one brick space. Insulation should not be allowed to enter the drainage cavity between the brick veneer and the stud walls.

  Remember that each stud space will require two or more holes with this method. The top hole should be no more than 300 mm (12 in.) from the top plate, preferably 150 mm (6 in.).

  Make sure the installer patches the holes section by section rather than leaving them all until the end. If not, you may end up with several rain holes in your wall when that flash storm hits.

- **From the basement/attic.** This can be the easiest approach as long as the cavity is open from top to bottom, such as with balloon frame construction. A long tube is inserted in the cavity from above or below to within 150 mm (6 in.) of the bottom or top of the stud space. The hose is then withdrawn, 300 mm (12 in.) at a time. At each stage, the space is allowed to completely fill with insulation.

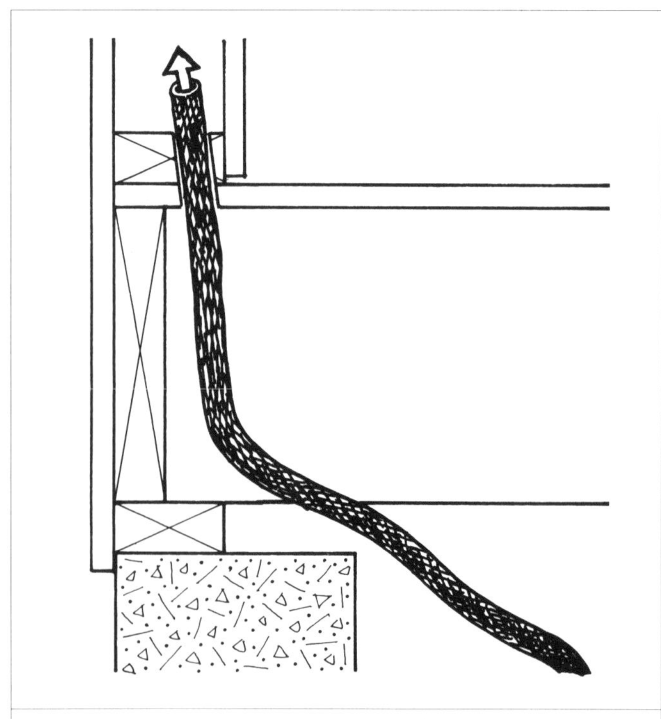

Blown-in insulation from the attic or basement leaves the walls intact.

All stud spaces in the wall need to be filled but there should be allowance for windows and doors, firebreaks, cross braces, and other obstructions in the wall cavity. The contractor should first plumb the space and verify the horizontal spacing of studs.

Potential blockages in the wall include wires, pipes, blocking, and windows and doors.

Just as you should choose your contractor carefully, carefully choose the insulation to be used in consultation with the contractor. The characteristics of the different loose fill insulations are outlined in Chapter 2.

Note: Blown-in cellulose fibre will more readily fill irregular spaces than the other materials. Cellulose is also the only blown-in insulation that can significantly restrict **air flow** when blown to proper densities. Stipulate in the contract that the density should be no less than **56 kg/m^3** (3.5 lb/ft.^{3}). This density is approximately 1.5 times the density of insulation normally used for attic applications.

When you have chosen the material, figure out with the contractor how much should be used. Knowing the size of the wall to be filled and the density of the application, you and the contractor should agree on the number of bags to be used — **and write it into the contract**. Only a small variation from the target is acceptable. If the contractor uses too little, the insulation settles and leaves gaps in the wall. If too much is used, some of the insulation may be escaping from the wall into a floor space or some other area where it is not needed — a big waste! Make sure the right amount is used.

Remember to tighten up the air barrier system. Seal all air leaks into the wall and keep your humidity levels low. A coat or two of low permeability paint (oil-based or latex vapor barrier paint) applied to the inside surface of the wall provides a vapor barrier.

Part II Renovating the Interior

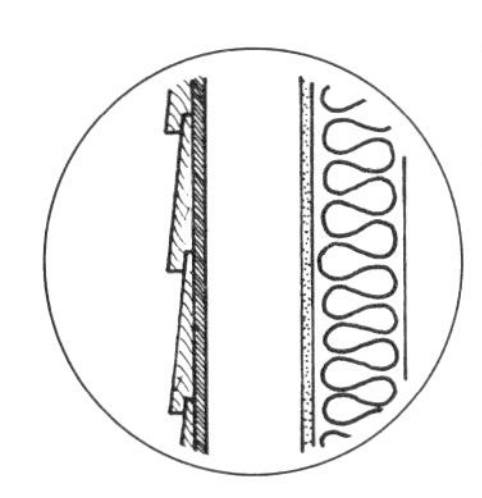

If your plans involve extensive renovation, you have two options:

- **Rebuild the existing wall.** If you have a wood frame house and you are removing the existing wall board or plaster, you can easily insulate the cavity. You could save even more energy by adding insulation to the inside surface. Attach 38 mm x 38 mm or 38 mm x 64 mm (2x2 or 2x3) strapping across the studs to allow for a second layer of insulation in the wall. Alternatively, you can mechanically fasten (not glue) rigid board insulation directly on the exposed studs. High R-value boards are more expensive but thinner, using up less interior space.

- **Build a new wall on the inside of the existing one**. With both wood frame and masonry walls, you may build a new wall inside the existing one, and then insulate it. If the old wall has an existing vapor barrier, such as polyethylene sheet, it's preferable to remove it. If the vapor barrier is in the form of an oil-based or latex vapor barrier paint, you can reduce its effectiveness by scraping the surface. The new vapor barrier should be more effective. Typically, a polyethylene sheet is installed as a combined air and vapor barrier. The new frame wall can be installed any distance from the old wall, depending on the level of insulation desired.

In both cases, follow the relevant parts of the instructions for insulating a basement from the inside. This includes: sealing air leaks; framing the new wall; framing around window and door openings; insulating (ideally in two layers so there are no gaps); installing an air and vapor barrier; and installing new drywall. A few additional points should be noted:

- If there is an existing uninsulated cavity it may be easier and less expensive to blow-in insulation rather than expose the cavity first.

- When installing a polyethylene air and vapor barrier, unroll the sheet across the entire wall area, **including** window and door openings. Cut these later with an "X" from corner to corner and seal the flaps to the frame. Make sure the air and vapor barrier is well sealed at all joints, openings and interruptions.

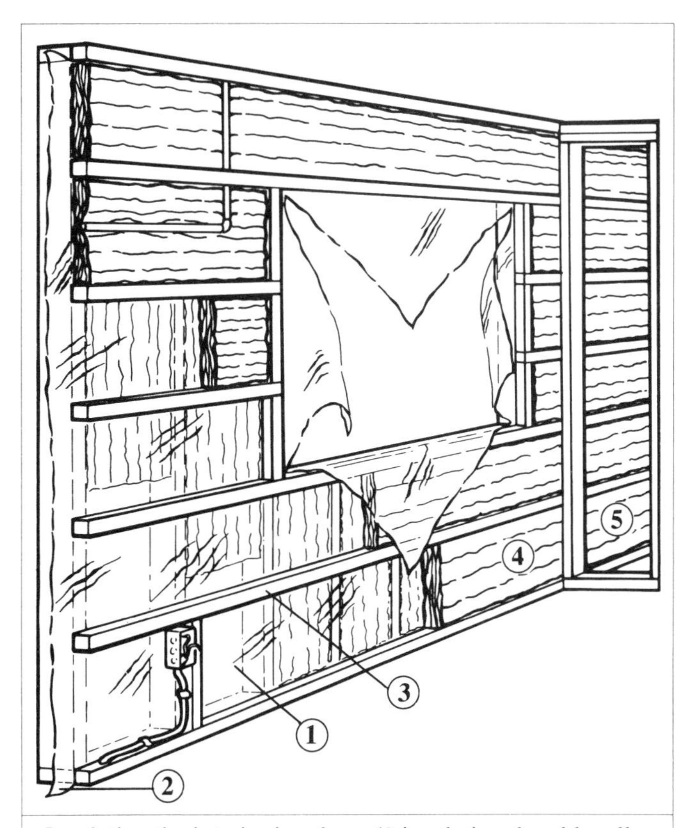

Insulating the interior involves: 1) insulating the old wall; 2) applying the air and vapor barrier; 3) cross-strapping; 4) horizontal insulation between the strapping; 5) option of extending the insulation past partition walls.

Caution: Two-thirds or more of the total insulating value of the wall must be on the cold side of the air-vapor barrier. Refer to page 17 for more details.

- Before you start the work, identify services within the wall. These will include plumbing, wiring and heating and ventilation ducts. Plan for the insulation to extend behind any pipes, electrical boxes, and so on, so that these obstructions are on the **warm** side of the insulation. This may be difficult when building new inside walls. Electrical boxes may be moved into the new wall. Pipes can be more difficult. If there are any in the existing wall (follow them up from the basement and down from the bathroom and kitchen), they could freeze and burst if left **outside** the insulation. Move the pipes or get insulation **behind** them, rather than in front.

- If you are rebuilding the existing wall, seal all cracks around door or window frames with polyurethane foam sealant, or stuff the gap with insulation and caulk. If installing a new wall, you must frame the windows and doors.

- For non-standard stud spaces, cut the insulation (if the batt type is used) about 25 mm (1 in.) wider than the space to be filled.

- Consider extending the insulation past junctions at partition walls. To do this, remove the drywall on the partition wall back one stud space and cut back the partition wall enough to extend the insulation and air and vapor barrier past the partition wall.

- If space is at a premium, insulate with rigid board insulation. This will cost more for the materials, but will require less space for the same insulating value.

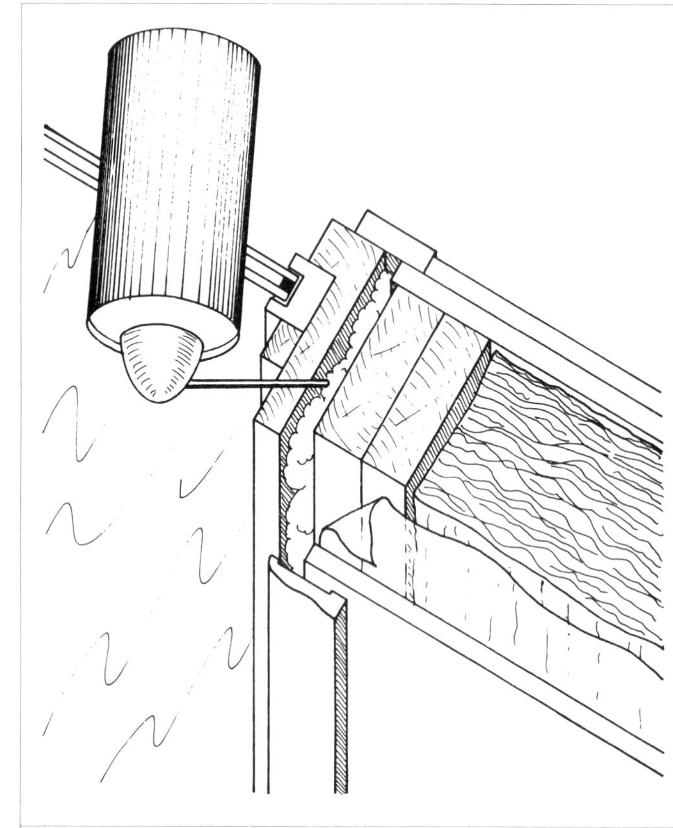

Foam insulation, used carefully, can insulate and seal the gap around window and door frames.

Part III Renovating the Exterior

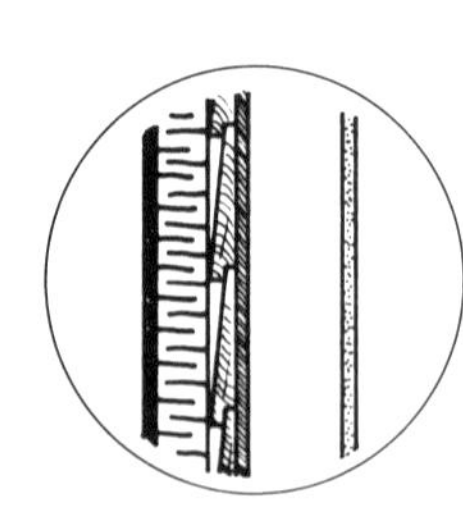

Installing new siding on the house provides an opportunity to install a rigid board insulation or batt/ blanket insulation over the old siding, but under the new. A few points to keep in mind:

- It is possible to add significant amounts of insulation to the outside since space is usually not a limitation (except in the case of lot line restrictions).

- Insulate any uninsulated stud spaces in the existing wall before adding to the exterior.

- Make allowances for extending window and door jambs, and for other penetrations such as vents, electrical service, gas and oil pipes, and for eavestroughs and downpipes.

- Consider the location of the vapor barrier. If the new insulation has at least twice the insulating value of the old wall, then a new air and vapor barrier can be installed over the old wall, before installing the new wall and insulation. This is most often the case with solid masonry walls, and can be done with a continuous and well-sealed polyethylene sheet.

- Make sure there is not an air space behind the existing siding which would allow cold air circulation and short-circuit the new insulation.

- Don't forget air leakage control at penetrations, seams, and edges, especially at the top/eaves area. This is to ensure that house air doesn't circulate through the old wall and out at the eaves. Seal penetrations and leakage points on the interior wall.

- If your basement is not already insulated, consider extending the insulation below ground level over the foundation wall (see Chapter 5 for details).

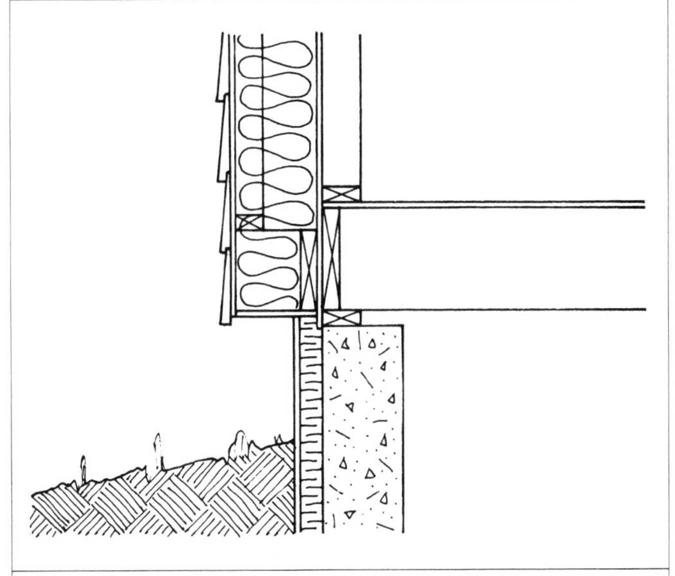

Extending the insulation below grade provides continuous coverage.

- The framing generally requires a top plate along the soffit and a bottom plate attached to the foundation wall. In some cases it may be necessary to go up into the eaves (not a problem if you were planning on replacing the soffits anyway).

- Ensure that the eaves will prevent water from getting in between the insulation and the siding. If necessary, add flashing at the top of the insulation. Caulk the top joints for good measure.

Rigid Board Insulation

- Any type of rigid or semi-rigid board insulation can be used.

- Fasten rigid insulation in place, with the appropriate nails and washers; check with the manufacturer or supplier.

- Make sure that the framing and insulation fit snugly together without gaps.

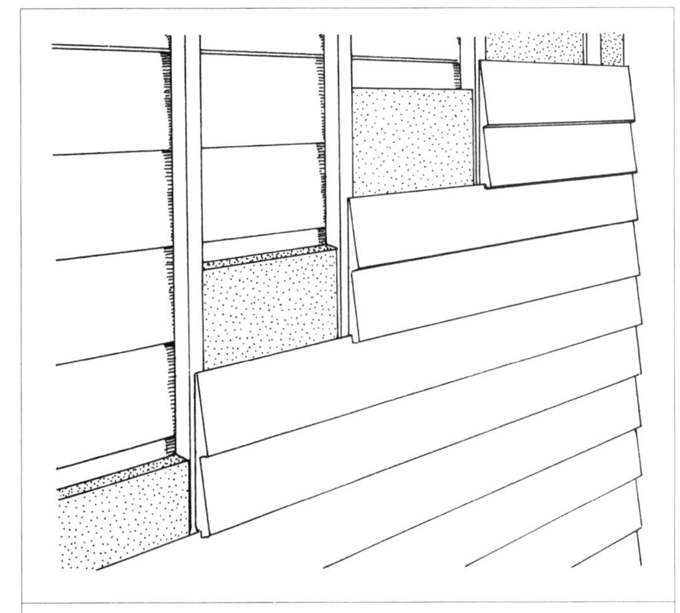

Strapping and insulation should fit snugly together.

- The new cladding should use fasteners long enough to penetrate the nail base by 30 mm (1 1/4 in.).

- Insulated siding (often polyurethane foam sprayed onto the back of aluminum or vinyl siding) is an alternative to separate insulation and siding. The insulating value is not high, up to RSI 0.70 (R4). It should be installed so that there is no air circulation between the new siding and the old wall, which can be very difficult to accomplish.

- Strapping **over** the insulation rather than on the side of the insulation has several advantages. It provides a rain screen and reduces heat buildup behind the siding. It also provides a nailing surface for the siding without creating a thermal bridge.

Batt/Blanket Insulation

- Build a wooden framework over the entire outside wall to hold the insulation and support the new siding. Cross strapping, in layers perpendicular to each other, can build out the new wall. Alternatively, a lightweight frame wall can be hung from the rafters or supported on a bottom plate out from the old wall. This would allow two layers of insulation, one horizontally behind the frame, the other vertically between the studs. In this way RSI 3.5 (R20) or higher can be installed.

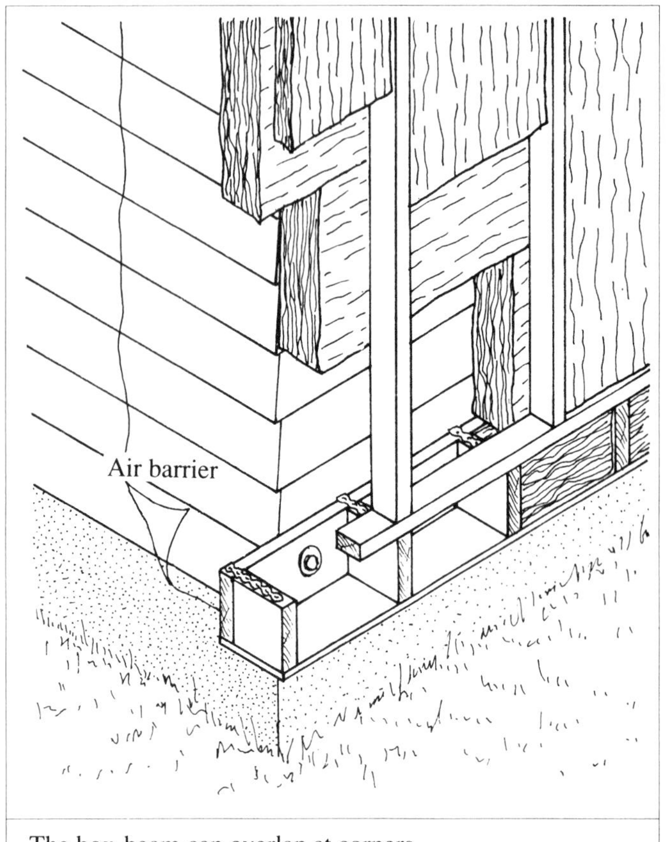

The box-beam can overlap at corners.

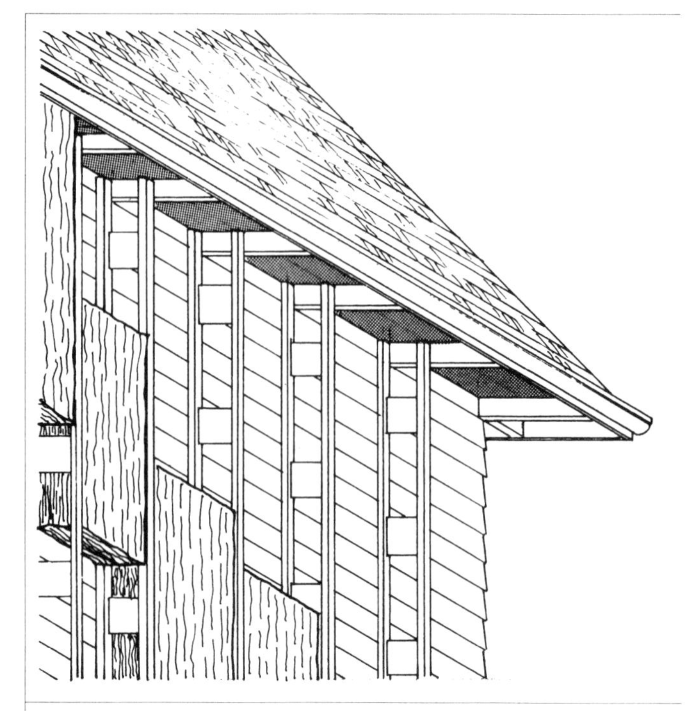

Homemade trusses are hung from the rafters and nailed to the existing wall.

- Batt/blanket insulation should fit snugly into the supporting framework without gaps.

Part IV Miscellaneous Spaces

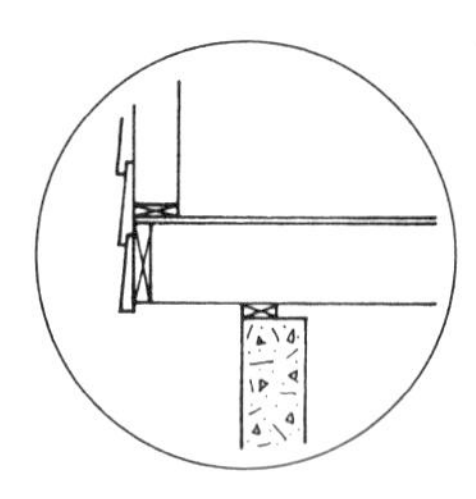

Everyone knows that warm air rises. But heat radiates in all directions, so walls and floors must be insulated where they separate a heated space from an unheated space.

This section deals with three special cases: unheated garage, cold cellar, and overhang.

Unheated Garage

The walls and the ceiling adjoining the house must be insulated.

The Walls
Refer to Parts I and II of this chapter for instructions on wall insulation if the garage is above ground; refer to Chapter 5, Basements, Parts I and II, if it is below ground level.

The Ceiling
If the ceiling is open and the joists are visible, proceed as outlined for open foundations in Chapter 5. It may be desirable and relatively easy to remove an existing ceiling finish.

A contractor can blow the ceiling full of insulation. Read the comments in Chapter 4, Part III on how to insulate a covered attic floor. Any holes cut in the ceiling should be carefully re-sealed to prevent gas fumes from vehicles leaking into the rooms above. Insulation must be blown to the proper density to prevent settling. Cellulose insulation will provide additional air leakage control if applied to the right density.

If there is a ceiling finish that you do not want to remove, you can nail rigid board insulation to it as long as the surface is fairly even. The rigid board may require a fire protection layer depending on code requirements. Seal any potential air leakage paths that would allow air to bypass the insulation. This is especially important at the edges where it may be necessary to expose around the perimeter of the ceiling. Seal this area with urethane foam, or with sections of impermeable rigid board insulation caulked between the joists.

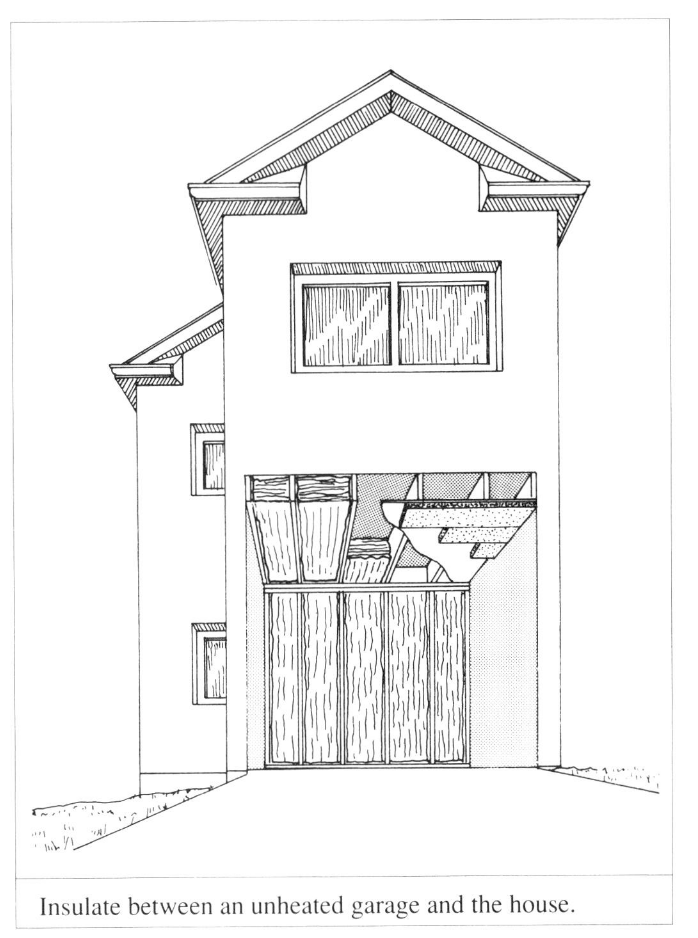

Insulate between an unheated garage and the house.

Cold Cellar

The cold cellar walls (separating the heated basement from the cold cellar), door and ceiling should be insulated. Apart from saving heat, your cold cellar will be colder.

Treat the walls as outlined in Chapter 5, making sure that the air and vapor barrier is on the warm side.

Treat the ceiling as outlined in the section on unheated garages above.

Overhang

Seal the floor of any overhang and fill it with insulation.

- If the overhang is on the first floor, this might be accomplished from the basement by stuffing in some batt insulation to fill the space. The air and vapor barrier should face up to the warm side. See Page 72 for more detail.
- If the overhang is between the first and second floor, the problem is more difficult. If you can easily lift the floorboards or remove the outside surface, fill the space with batt insulation. Seal any openings into this space.

Contracting the work

Generally, blowing insulation into a wall is best done by a contractor with specialty equipment. Include these details in the contract: areas to be insulated, access details (where the holes are drilled, size and spacing), type of insulation used, density of blown-in insulation, number of bags of insulation, clean-up and finishing (final finishing of the hole plugs is often the homeowner's responsibility).

Look for a contractor who discusses the need for air sealing and a vapor barrier on the warm side of the insulation, and who suggests methods of sealing some of the more hard to get at locations, such as overhangs.

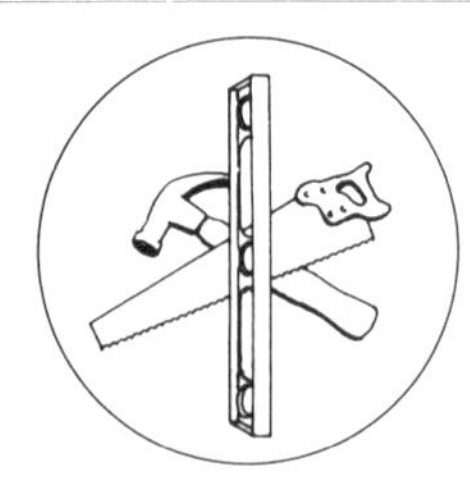

Part V Additions and New Construction

Renovation activity often involves some new construction, such as the addition of a room or wing. New construction provides an opportunity to install a continuous air and vapor barrier and high levels of insulation in an efficient and cost-effective way.

The illustration shows a typical cross section of new construction from the roof to the footings. Note how both the insulation and air barrier run continuously without breaks or thermal bridging.

Attic:
High levels of insulation, a continuous air and vapor barrier, and ventilation are the features of an energy-efficient attic. Roof trusses are available that allow high insulation levels over the top plate of the outer walls. These include the dropped chord truss, and scissors and parallel chord truss for cathedral ceilings.

Walls:
The section in the illustration shows a 38 mm by 140 mm (2x6) wall with insulating sheathing. Other systems include interior cross-strapping, double-wall systems, and the use of trusses. These systems allow the continuous air and vapor barrier part way in the wall. Note the recessed headers which allow the continuous air and vapor barrier and extra insulation.

Windows:
High-performance windows are used wherever possible. This means double-glazed with a low-E coating, or better. The majority of windows face south. All windows are sealed to the air barrier.

Foundation:
The foundation has full-depth insulation, in this example, on the exterior. Proper damp-proofing, a drainage system and sloped grade help ensure a dry basement.

Mechanical Systems:
Heating requirements are less than for a conventional structure. Combustion air will be needed for all fuel-burning appliances — or better yet, use appliances that require little or no household air to operate. Refer to Chapter 8 for more information on mechanical systems.

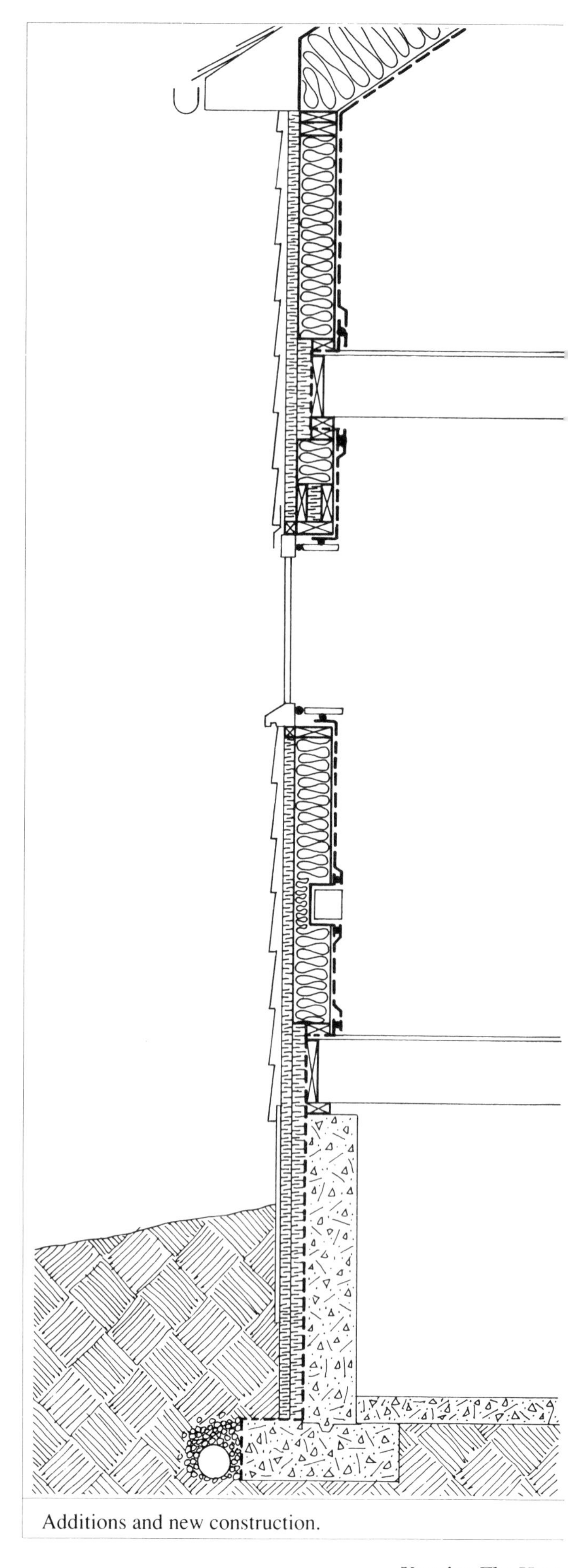

Additions and new construction.

7 Upgrading Windows and Doors

Part I
Windows

Part II
Doors

Introduction

Windows and doors can be big energy wasters for three reasons:

- Glass itself is a highly heat-conductive material. Similarly, many wooden doors are highly conductive.
- Doors and operable windows have many paths where air leakage can occur.
- Air can pass through the joints around window and door frames unless they are sealed tightly.

This chapter deals with weatherstripping and other ways of upgrading windows and doors to save energy.

Part I Windows

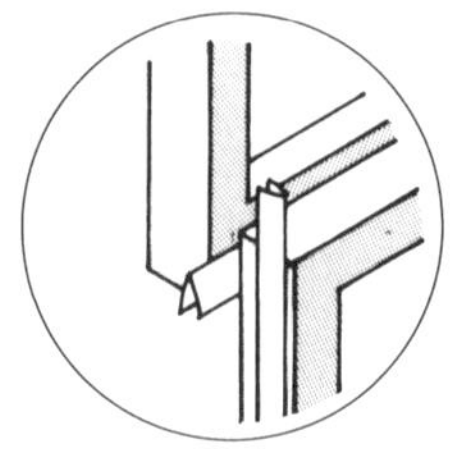

Taking Stock

Check the window for signs of damage before starting any work. This includes checking for rot, mold, and staining on or around the window, the condition of the glass, putty and paint, the type of operation, and the condition of the closing hardware. Some repair may be needed.

Check the weatherstripping at all movable joints. Combine a visual inspection of the existing weatherstripping with a test using a leak detector as described in Chapter 3, Air Leakage Control. New weatherstripping is a good investment in comfort and energy savings.

Check the glazing. Windows should be at least double glazed. There are many ways of adding extra glazing from the inside or outside, permanently or seasonally.

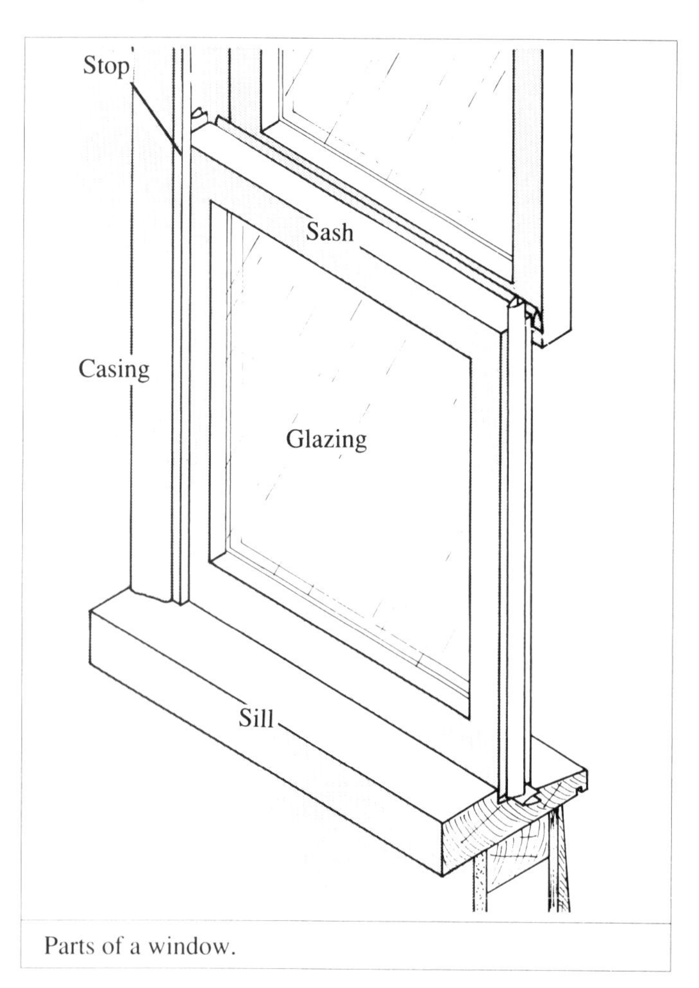

Parts of a window.

Consider new windows. If some of your windows are beyond repair, or you are planning to install new windows, look at some of the higher efficiency models. Energy savings alone rarely justify buying new windows. However, if you are planning on new windows anyway, your choice of the type of window can make a difference in fuel savings.

Condensation Problems

Condensation and frosting are common window complaints. Sometimes the problem is light fogging of some windows, other times there may be persistent and heavy frost covering the glass. Many homeowners buy new windows, only to find the problem has become worse.

Condensation occurs when water vapor in the air is cooled to the point where it condenses on the cold surface as water droplets or frost. The more humid the air and the colder the surface, the greater the accumulation of condensation. One solution is to reduce the humidity levels in the house as discussed on page 95.

Alternatively, you can increase the surface temperature of the window and frame. This can be done by having at least two layers of glazing. Better yet are windows with a "Low-E"-type coating or windows that are triple-glazed. Frames should be relatively non-conducting, such as wood, fibreglass, or vinyl. Metal frames, even if they have a thermal break, may be more susceptible to condensation.

Often the condensation is **between** panes. This is a result of moist house air leaking past the first pane and condensing on the cold surface of the outer pane. Even dry houses can suffer from this type of condensation problem. This problem is common on second stories where there is more air being pushed out the window because of the stack effect. The solution is to weatherstrip the inner sash to prevent air leakage; make sure that the weep holes are open to the outside.

If condensation occurs inside a sealed double-glazed unit, the problem can be corrected only by replacing the unit. Check to see if the window is still under warranty.

Weatherstripping

Windows should be weatherstripped around the sash to reduce air leakage. If the windows don't have to be opened and don't serve as emergency exits, they can be locked and caulked.

When sealing windows make sure that the inside window is **sealed more tightly** than the outside window. Otherwise, you will encounter condensation problems as the warm, moist air from inside the house becomes trapped between the panes.

Many different types of weatherstripping are available. Chapter 2 lists a few of the more common varieties, but it is certainly not complete. Try to visit a supplier who stocks a wide variety of weatherstripping types. The cheaper types are usually less durable and less effective — don't choose just on the basis of cost!

In most cases, newer windows will already have ''built-in'' weatherstrippings, which can lose their effectiveness over time. Pry out a sample of the weatherstripping and take it to the window manufacturer or supplier for replacement with the same type.

The common "double-hung" and "single-hung" windows should be weatherstripped on the sides, top and bottom of the moving sash, as shown in the diagram. If there are drafts around the fixed portion, they should be caulked.

Preparation and installation is important. This usually involves the following steps.

- Adjust and square windows that are out of alignment.
- Remove old weatherstripping, caulking, and paint blobs. If the surface is very uneven, apply a bead of caulking under the weatherstripping or fill and sand the surface to make it smooth.
- Clean the surface with a clean cloth and fast drying mineral spirits or MEK (methyl ethyl ketone).
- Apply the weatherstripping. With doors and windows that are used often, you may want to reinforce the adhesive types with staples.
- Check the window for smooth operation.
- Periodically check the weatherstripping for wear.

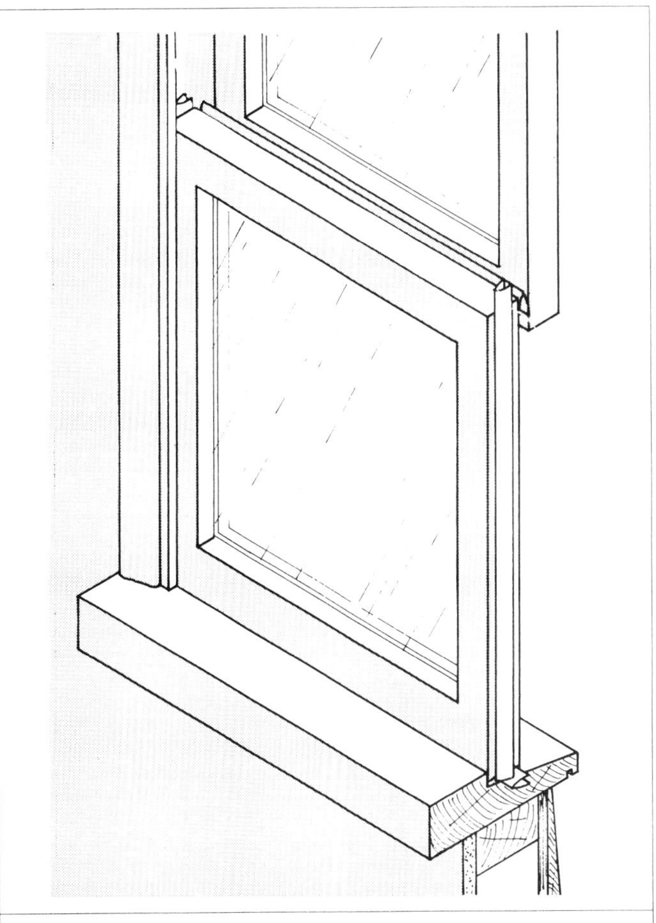

Windows require weatherstripping on all four sides.

Sides. The thin plastic V-type weatherstripping is a good choice. Open the window and slip the stripping up the crack between the sash and the frame, with the mouth of the "V" facing outside. It need only extend to 25 mm (1 in.) above the top of the closed window. You can do a better job if you first remove the stop and the bottom sash.

Top. Weatherstrip the space where the two sashes meet by removing the lower sash and applying V-type weatherstripping to the **upper** window from the **inside**.

Bottom. Apply V-type or compression-type neoprene rubber to the window sill where the closed window will sit, or to the bottom of the moving window sash itself.

Swing-type windows are treated like doors. Apply weatherstripping to the frame so it meets the edge of the sash, or place it on the stop where it will meet the face of the sash. The force of a closed window against the weatherstripping makes a pressure seal.

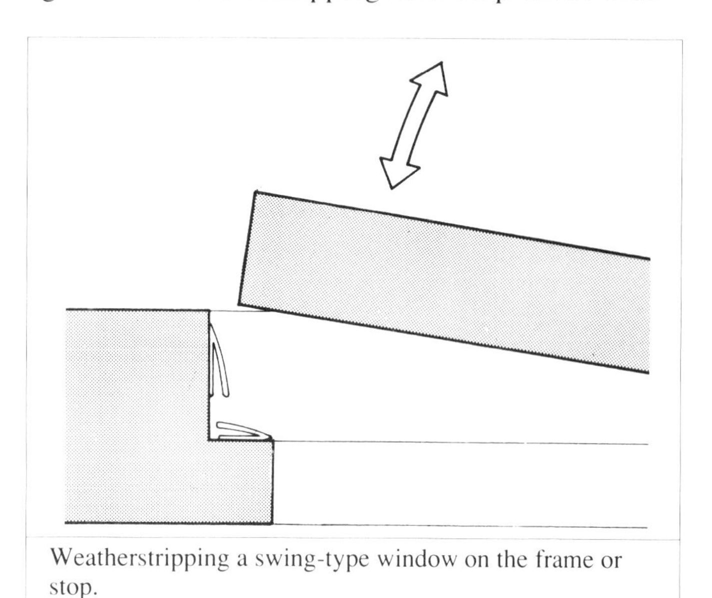

Weatherstripping a swing-type window on the frame or stop.

Single glazing has an RSI value of about 0.16 (R0.9), so it loses about 10 to 20 times as much heat as the same area of a properly insulated wall. Storm windows or double glazing will reduce the heat that is needlessly lost through the window in your house by almost half, while a higher efficiency window will reduce it by two-thirds or more! Improved windows will also make your house more comfortable by reducing drafts and increasing the temperature of the interior window, which would otherwise produce a cold feeling. Adding glazing also reduces the risk of condensation.

The range of the insulating value is determined by the number of panes, the thickness of air space, and any specialty coatings.

Window Products and Applications

Storm Windows

Storm windows can be installed on the inside or the outside and they can be permanent, seasonal, or temporary.

Polyethylene or Rigid Plastic Storm Windows

Installing plastic sheeting over the window is an inexpensive and easy way to improve the heat retention of your home and reduce condensation. The plastic will be less durable than glass and will have to be re-installed each year if the windows are opened. There is an inexpensive kit on the market that has a plastic sheet that is heat-shrunk in place using a hair dryer. This system is best for windows in seldom-used rooms and basements.

A do-it-yourself plastic storm window is available, made of a sheet of rigid plastic and a specially designed snap-in frame. Some types use a magnetic strip to hold the plastic in place. The mounting strip is permanently attached to the window frame, and the strip pops open to accommodate the plastic sheets. The sheet can then be removed and stored in the summer. The sheets must be handled carefully, because they scratch easily.

Interior storms must be well sealed to prevent condensation between glazings.

Note: Every bedroom must have at least one window that opens from the inside. This is required by the National Building Code for safety reasons.

Removable Storm Windows

Single-pane storm windows are designed to be installed each fall and removed each spring (unless you have air conditioning, in which case they stay in all year long). They can be made to order by suppliers listed under "Storm Windows and Doors" in the Yellow Pages. The major advantage of these removable storm windows over permanent ones is a lower cost.

These windows must be checked seasonally; damaged putty must be replaced and the frames painted regularly to preserve the wood. Conventional storms should not fit too tightly to avoid condensation between the storm and inner window. **The inner window must be very well sealed with caulking or weatherstripping.**

Permanent Storm Windows

The combination-type storm window has both screening and glass in the same unit, serving your needs year round. These windows have metal frames, and are made to order by specialized suppliers (under "Storm Windows and Doors" in the Yellow Pages). Installation can be done by the supplier or by you.

A permanent storm window can allow for ventilation.

These permanent storms are more convenient than the removable type, but are also more expensive and cleaning between the storm and the window may be a problem. Some units are available with an energy-efficient low-E coating. Regardless of which type you choose, shop around for well-made windows. Look at the quality of hardware and weather-stripping and strength of joints. These windows are a long-term investment and can increase your property value, so quality construction is important.

New Windows

Most existing windows can be repaired, weather-stripped, and have new glass added if necessary. However, there are circumstances when it is appropriate to buy new windows. New windows may be required when:

- the old windows have deteriorated;
- there is extensive damage (such as rot) to the surrounding wall caused by the old window;
- major renovations or additions require new windows; or
- you want changes in window size, operation, or the appearance of the house.

For further information on choosing new windows and doors, refer to page 107 to order a copy of the *Consumer's Guide to Buying Energy — Efficient Windows and Doors.*

Window Protection and Operation

Windows can be used to provide fresh air and remove excess humidity and odors. But be careful. It's hard to control windows as ventilators and it is easy to let too much cold air in.

Shutters, shades, and awnings can be used to keep the sun out when it's not wanted. Window insulation can be used in the winter to provide extra insulation for improved comfort and energy savings.

For window coverings to work properly, **they must fit tightly around the window frame**. Air passing behind the covering will increase heat loss and create a considerable condensation problem.

Curtains can help reduce radiant heat loss from windows.

Generally, the high cost of materials for insulating windows will make this a low priority item for most existing homes. If your windows need new shades or curtains anyway, consider methods for reducing heat loss at the same time.

One of the most practical solutions is to design or buy a **flexible insulating curtain** to pull across the windows or roll down from above. The curtain should ideally be made from a heavy, multi-layered material covered with your choice of fabric. Even one or two thick drapes over a window should do a good job if tightly fitted. The curtain should fit snugly along the window ledge, or be weighted for a tight fit along the floor. A valance or enclosure along the top is a good way to prevent air from falling behind the curtain.

Part II Doors

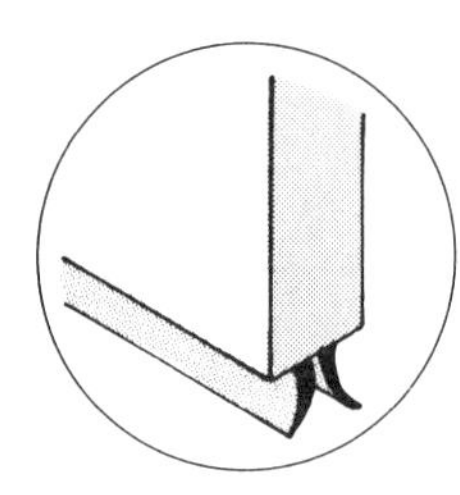

Doors, like windows, should fit snugly so the air can't sneak in around the edges. Poor installation, years of hard use, shifting foundations, and seasonal warping can often force doors out of line with their frames. If they don't fit snugly, fix or replace the weatherstripping.

The same techniques for preparing windows applies to doors. This includes any needed repairs or adjustments, surface preparation and cleaning, and fastening the weatherstripping.

Around the frame. Weatherstrip the tops and sides of any door **on** the frame, as illustrated.

The easiest and most effective weatherstripping for this use is a good quality V-shaped vinyl type. It makes contact with the **edge** of the door, and provides a good seal even when the door warps back and forth from season to season.

For increased protection, attach weatherstripping to the stop so that it presses against the **face** of the door as shown. There are also many types of combination metal and foam or rubber weatherstripping that are screwed to the stop. They should be adjusted regularly to conform to the changing warp of the door.

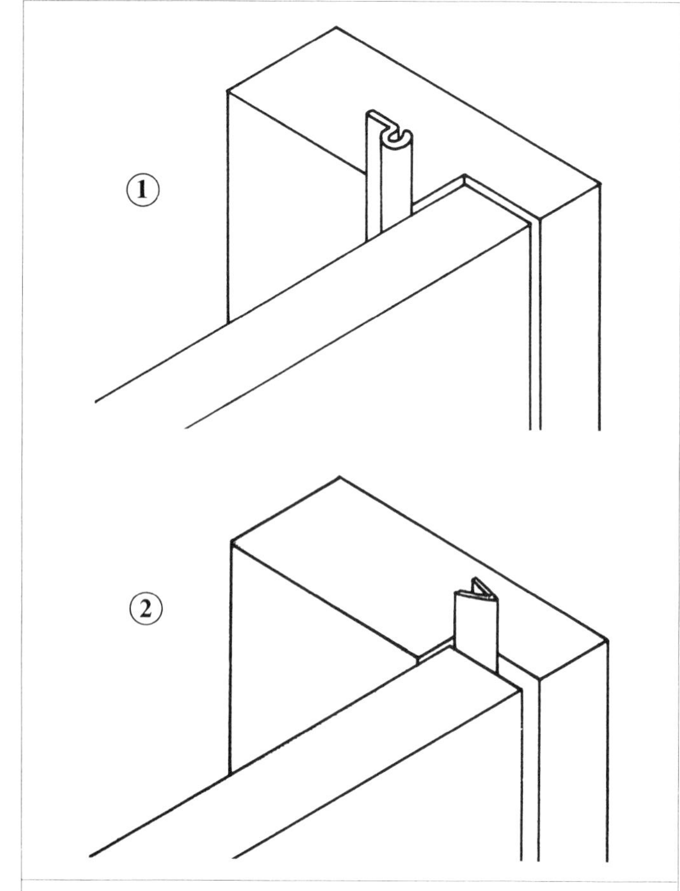

1) Weatherstripping the face of the door.
2) Weatherstripping the edge of the door.

Bottom of the door. Apply weatherstripping to either the door sill or to the door itself. This can be a difficult area to seal well, but it is worthwhile since this is often a source of major drafts.

Use durable material that can withstand the traffic and be flexible enough to conform to changes in the door caused by humidity and temperature. The weatherstripping should also be easy to replace. A good seal can usually be obtained with gasketed door bottom weatherstripping which attaches to the door, or with full or partial threshold weatherstripping which is attached to the door sill.

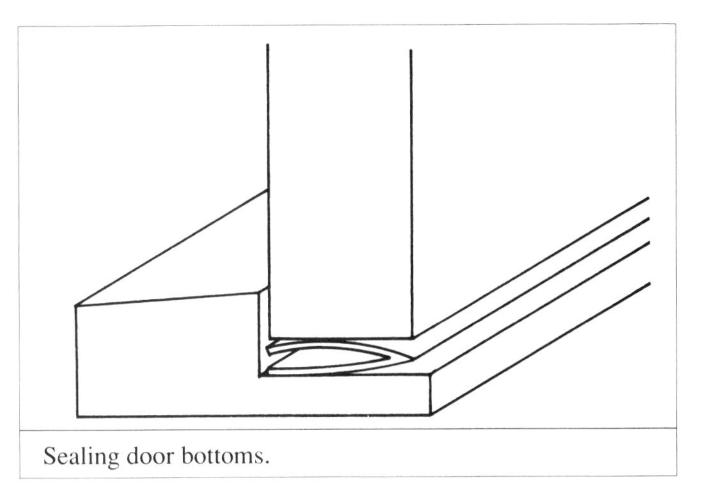

Sealing door bottoms.

When the weatherstripping is applied to the door itself, a very durable material is necessary. The most readily applied and effective choice is the combination type, which is simply tacked or screwed along the bottom **inside** surface of the door. There should be slots which allow for some adjustment of the weatherstripping.

Replacing the Door

Sometimes it is appropriate to buy new doors. Take this opportunity to buy an energy efficient one. A *Consumer's Guide to Buying Energy-Efficient Windows and Doors* can help you choose the right door for your home. Write to the address provided on page 107 to order your copy.

Contracting the work

Air sealing and repairing windows can be good do-it-yourself projects. Alternatively, these can become part of the work of an air sealing contractor (see Chapter 3). If you are hiring a contractor, consider asking for a quote on some related air sealing work.

If you are having some of the windows in your house replaced, ask the contractor to check the weatherstripping on the remaining windows, and replace it if it is worn.

8 Operating Your House

Introduction

By now you should be very familiar with the "house as a system" concept. Like any system, however, your house will only run as efficiently as it is operated. As the homeowner, this puts you in the driver's seat. How you live in your home and maintain and operate it will have a lot to do with the success of your retrofit activities. If you operate it poorly, you will negate a large part of your retrofit gains. If you operate it efficiently, you can actually improve heating performance. Even more importantly, you will create a healthier, more comfortable environment. Most furnaces and boilers operate below peak efficiency because of poor maintenance. A well tuned and efficiently operating heating system can significantly reduce your annual fuel bill.

Part I Operating and Maintaining the Heating System

Maintenance

Follow recommended maintenance procedures for cleaning and servicing. For an oil furnace and boiler this means a thorough cleaning and tuning each spring. For gas furnaces and boilers servicing should be carried out at least once every two years. The majority of the tune-up should be carried out by a qualified service technician or heating contractor. A series of booklets on heating systems (available from the address indicated on page 107) will give you all the facts you need about your heating system.

Homeowner Maintenance

If you have a forced air system, keep return air grills and warm air vents clean and free of obstructions, and change or clean your filters once a month.

If you have an electric baseboard system, keep your heating units clean. Vacuuming them twice a year should prevent dust build-up.

Hydronic systems perform best when radiators are relatively free of air. This means you must remember to bleed your radiators on a regular basis unless the system has an automatic bleeding capability. Further savings can be achieved by adjusting the operating temperature of the water to reflect outside conditions. A control can adjust the water temperature to match the changing demand for heat so you'll never be too hot or too cold, or pay for heat you don't need.

Thermostat

The greater the difference in temperature between inside and outside, the greater the heat loss. By turning your thermostat down a bit, and a bit more at night, you'll lose less. The best way to ensure that the thermostat gets turned down is to install a clock or setback thermostat. The setback thermostat will perform predetermined temperature setbacks automatically.

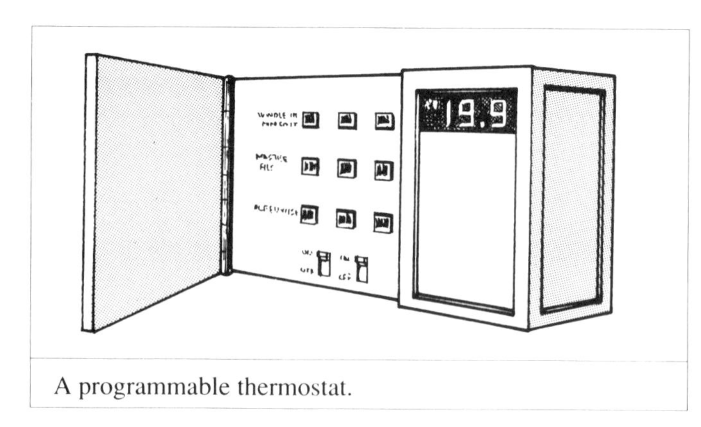

A programmable thermostat.

A good quality programmable thermostat will provide a clock timer and allow you at least two setback and reset periods a day. For example, a temperature reduction could be programmed in to start before you go to bed and end before you get up in the morning. The second setback can reduce house temperature when everyone is out of the house during the day and end just before you arrive in the evening.

Sizing and Balancing Heating Systems

A newly retrofitted house will have a smaller heating need than it had before the retrofit. A smaller heating requirement means that the furnace is oversized, a condition that can cause larger temperature fluctuations and the potential for inefficient on-off cycling of conventional furnaces.

When a furnace operates less often, the chimney can get colder between burns. This can lead to downdrafting, and possible condensation and damage to the chimney. If your retrofit measures are extensive and you have a concern about this, have the system checked by a qualified heating system technician.

In addition, the balance of the heat distribution will likely have changed. Insulation and air sealing could mean that some hard-to-heat rooms are now easier to heat, while other rooms may overheat. This may require a re-balance of the system by either adjusting the dampers in a ducted system, or adjusting valves in a hot water system.

Ask the service mechanic or heating contractor who maintains your system what changes may be required so that your system operates at top efficiency.

Heating Ducts

Heating ducts running through unheated or cool basements should be insulated. First tape the joints to prevent leakage. While you're at it, it would be well worth your while to tape the joints of all heating ducts regardless of whether they are to be insulated or not. Then take 75 mm (3 in.) or more of mineral-fibre batts or blankets and cut them to size. Specially designed batts are available for this purpose. Wrap the insulation around the ducts, or else install it lengthwise. Make sure the entire system is covered, with the insulation taped securely. Don't wrap the ducts of a wood-fired furnace within 1.8 m (6 ft.) unless you use a special **non-combustible** insulation designed for wrapping ducts.

It is also a good idea to check return air ducts, especially where they pass through crawl spaces or garages, and caulk or tape any leaky joints.

For those with hydronic systems, placing a foil covered insulation board between radiators and exterior walls will reflect heat back into the room that would otherwise be lost to the outside.

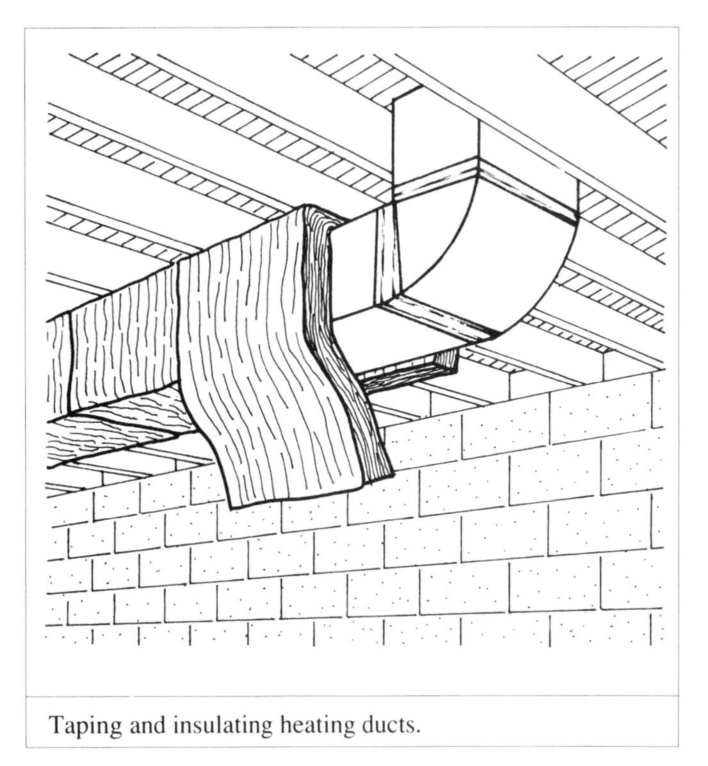

Taping and insulating heating ducts.

Fireplaces

The heat supplied by a fire will not make up for the loss of heat when house air is drawn up the fireplace chimney. Consequently, most fireplaces are unable to provide any net heat gain. There are several fireplace accessories such as tightly-fitted glass doors and outdoor air intakes that offer nominal improvements on fireplace efficiency.

For information on how to improve fireplace performance read *A Guide to Residential Wood Heating,* which can be obtained from the address listed on page 107.

Domestic Hot Water

Domestic hot water systems also consume large amounts of energy — more than all your lights and appliances combined. In fact, **next to space heating, the water heater is the largest energy user in the home**.

An economical way to reduce domestic water heating costs is to wrap the storage tank with additional insulation. An **insulation kit** for your hot water heater can be purchased from hardware stores; it contains a vinyl-covered insulation jacket, pre-cut tape and instructions. Use only insulation blankets that meet the latest edition of the Canadian General Standards Board (CGSB) specification 51.65M, "Insulating Blankets for Domestic Hot Water Heaters". This specification is currently in the process of becoming a standard.

Follow the manufacturer's instructions when installing kits.

According to the CGSB specification 51.65M, you may cover the controls of an electric hot water heater as long as they remain accessible.

When insulating a **gas water heater do not cover the controls or obstruct the vent connections or combustion air openings**.

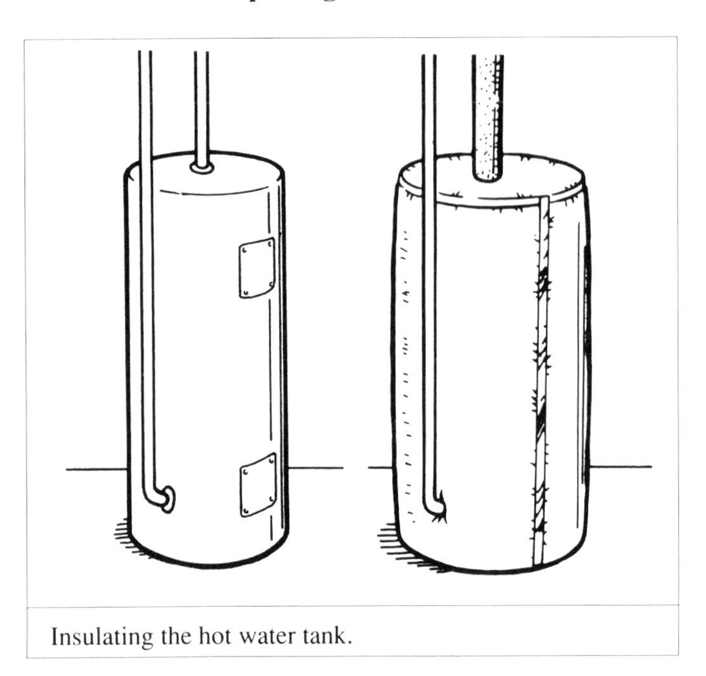

Insulating the hot water tank.

For more information on insulating hot water tanks, consult with your local fuel supplier.

Insulating the hot water tank should save enough money to pay back the costs of materials quickly. You can make even greater savings in your hot water bill by acting on the suggestions that follow.

- Insulate hot water pipes that pass through unheated areas or service washing machines. The insulation should be at least 13 mm (1/2 in.) thick to be effective. You can use pieces of batts with duct tape, or the wrap type insulation, but the ideal pipe insulation for existing homes is the snap-on type. These are long tubes slit lengthwise that snap over the piping and then are glued shut.
- Install a flow restrictor in the showerhead pipe and aerators in the faucets. These inexpensive gadgets will allow you to do the same job with much less water.
- Fix leaky faucets. At 1 drop per second, you're wasting 720 litres (160 gallons) per month — or 16 hot baths!
- Run washing machines with full loads, and rinse with cold water.

Part II The Heating System and Ventilation and Combustion Air

Introduction

Both you and your furnace need some fresh air coming into the house. Most Canadian homes, though, have too much. In fact, 25 per cent or more of your heat loss can be due to excess infiltration around windows, doors and other cracks. These drafts not only cost you money, but can make your home unpleasant during the winter.

How tight is too tight? When do humidity and condensation become problems? What are the requirements for ventilation and combustion air? This section looks at some of the implications of air leakage control: how it can affect air quality and air needed for combustion appliances.

It is worthwhile to take a systematic look at the moisture balance and ventilation needs of your house. This will involve a list of moisture sources, symptoms of problems, and ventilation requirements. Consider how any renovation or retrofit plans will affect the house. For example, if the house already shows signs of excessive condensation, any plans that include making the house more airtight will have to include increased ventilation.

For many, if not most, older homes a program of comprehensive air leakage control will not reduce the air supply enough to cause problems. Most older houses experience too much air leakage, even after air sealing.

Exceptions

However, there are times when problems can occur, even without thorough air leakage control. It is important to be aware of the potential problems, the symptoms to look for, and some of the possible solutions. These circumstances can make a house more susceptible to problems:

- houses with electrical heat or a high efficiency furnace (no conventional chimney);
- houses where there is competition for air (fireplaces, powerful exhaust vents);
- houses where there are sources of air contamination (smokers, hobbies, etc.); and
- houses that produce a lot of moisture and have high humidity levels.

Symptoms of Problems

If you are aware of the symptoms indicating the start of a problem, you can make adjustments and correct the situation before damage results. Some of the signs to look for include:

- excessive condensation on double-paned windows;
- staining and mold growth, which often appears first in bathrooms, closets, and on walls or ceilings at corners;
- stuffy atmosphere and lingering odors;
- back-puffing/odors from the furnace; and
- backdrafts from the fireplace.

Solutions

If the problem is one of high humidity or condensation, the first step is to reduce the humidity level by controlling the amount of water vapor that goes in the air. The following suggestions will all reduce the humidity levels in the home.

- Don't store wood in the house.
- Avoid hang-drying laundry in the house.
- Disconnect any humidifiers.
- Cover any earth floors in basements or crawl spaces with a moisture barrier.
- Install a sump pump to remove excessive moisture from the soil under the slab.
- Fix all water leaks into the basement.
- Don't allow any standing water in the house or against the foundation wall.
- Make sure the ground slopes away from the foundation wall and that there are properly-functioning eavestroughs around the house.
- Ventilate kitchens and bathrooms during use.
- Watch your living habits to produce less humidity (cleaning, washing, cooking, etc.).

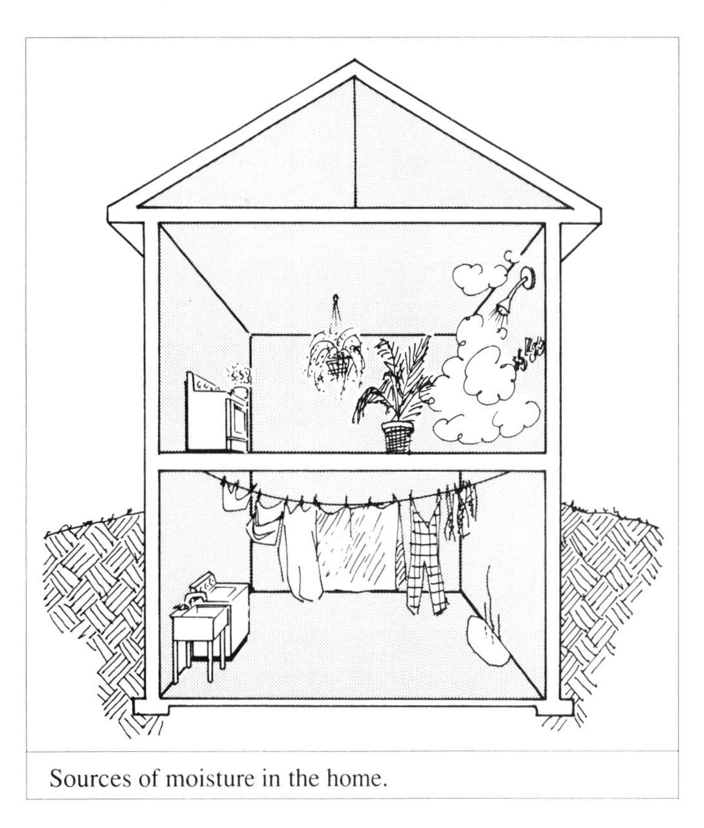

Sources of moisture in the home.

This table presents the maximum levels of humidity at which there will be no condensation on double-glazed windows at various outside temperatures.

Outside Air Temperature (°C)	Maximum Inside Humidity at 20°C (68°F)
– 30 or below	15%
– 30 to – 24	20%
– 24 to – 18	25%
– 18 to – 12	35%
– 12 to 0	40%

It can be very difficult to accurately measure and maintain the recommended humidity levels. A simple approach is to let the house become your indicator. If condensation starts to appear on the indoor face of any double-glazed windows (except those in the kitchen and bathroom), you have found the balance point. Occasional condensation does not pose a problem. Excessive condensation or frosting is an indication that you should reduce moisture production or reduce humidity levels by increasing ventilation.

Increasing Ventilation

If you still have too much condensation even after reducing moisture production, or your problem is one of air quality, you will have to increase ventilation. There are several ways of doing this; the best method will depend on your house, the degree of your problem, and your preferences.

Kitchen and bathroom fans should be used regularly.

- A simple solution may be to turn on kitchen and bathroom fans more often, especially when those rooms are used. A simple timer switch will turn the fan off automatically to prevent excessive ventilation.

- If you don't have ventilation fans, have them installed. It is worthwhile buying quieter fans since they tend to be used on a regular basis. Make sure that the fan installation incorporates air-sealing measures.

- Advanced, high-efficiency combustion appliances are ideal for air-tight houses. This creates far fewer problems than conventional appliances, and at the same time, increases energy-efficiency benefits.

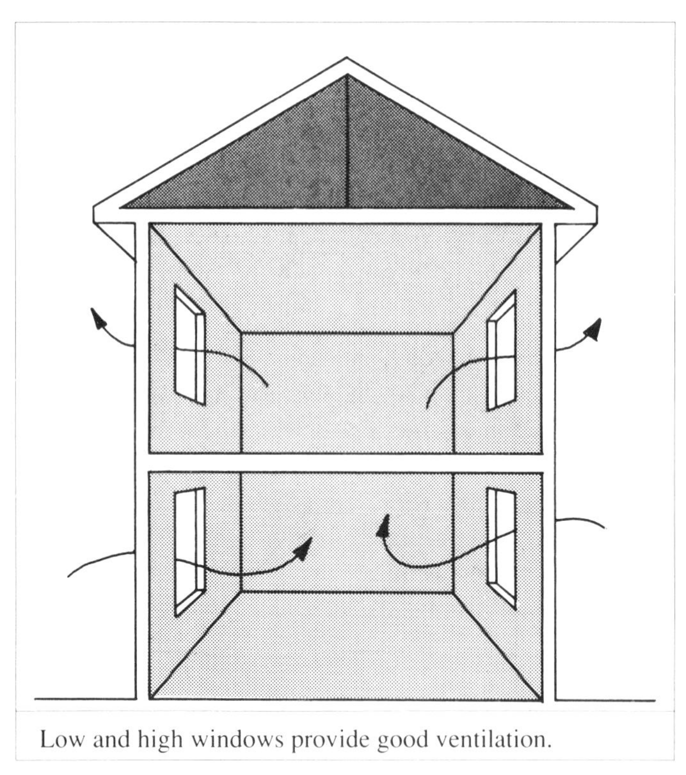

Low and high windows provide good ventilation.

- Open windows as needed. Windows open on either side of the house or on different floors will help ensure better circulation. It is harder to control the rate of ventilation by this method.

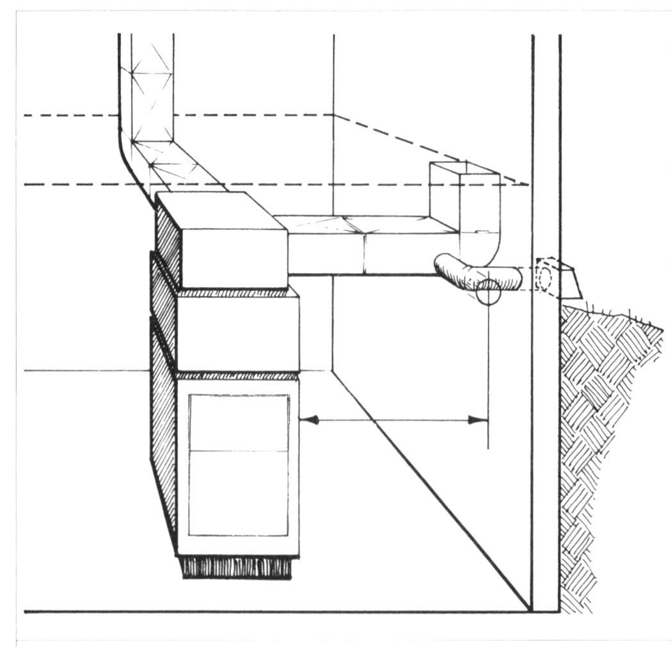

Fresh air duct to the cold air return.

- If the need for ventilation air supply and distribution is great, a central ventilating system can be installed. A relatively inexpensive and effective technique involves connecting a fresh air duct to the return air plenum of a forced-air system. Fresh air is drawn in by the suction of the furnace fan, mixed with house air, and pre-heated by the furnace. Open the damper in the fresh air duct just enough to prevent window condensation. It will have to be adjusted periodically through the winter. Alternatively, you can install a motorized damper operated by a humidistat control. It will open the damper only when the house becomes too humid.

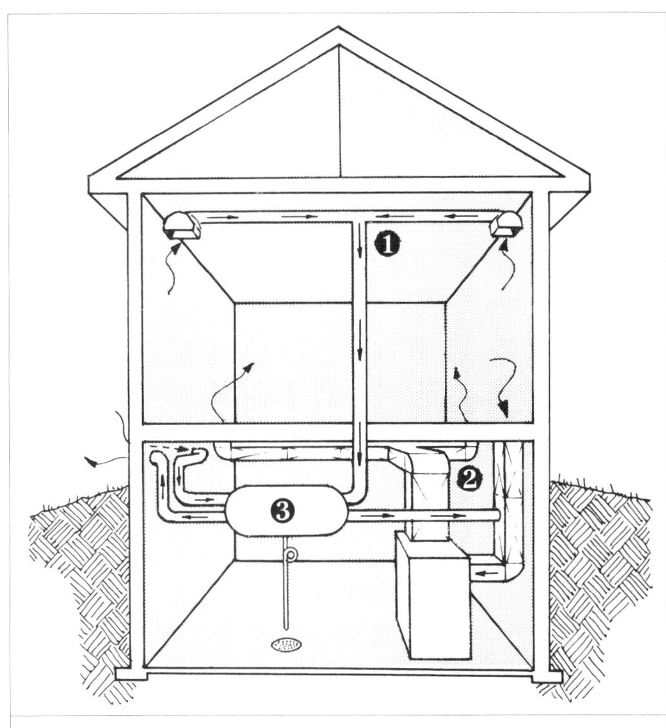

Heat recovery ventilators consist of: 1) collection and exhaust of stale, moist air; 2) supply and distribution of fresh air; 3) heat exchanger to recover some of the heat from the outgoing air.

- There are other types of whole house ventilation systems that include exhaust and/or supply fans. These systems are for very tight houses that require a great deal of ventilation. The air must be distributed throughout the house and in a way that maintains comfort levels. The system should not create strong negative or positive pressures in the house. This means that the inlet and outlet air supply has to be in balance.

- Some systems are designed with the exhaust fan in the attic and several ducts pulling air from the kitchen and bathrooms. Many central ventilation systems use a heat recovery ventilator that typically recovers 70 per cent of the heat from the exhaust air and transfers the heat to the incoming air or to the hot water supply. Central ventilation systems should be designed and installed by a professional.

Combustion Air

Furnaces, fireplaces, wood stoves, and any other fuel-burning appliance also require air for combustion and for diluting and exhausting the products of combustion out of the house. **If there is not enough air, it is possible that the chimney or flue could backdraft, or spill dangerous gases into the house.**

Backdrafts or spillage may be caused by competition for air. For example, a roaring fireplace, a powerful kitchen ventilator, a barbeque range, or even a clothes dryer vented to the outside can exhaust a lot of air from the house. If the house is too tight, this can cause air to be pulled into the house through the chimney or vent, resulting in backdrafts or spillage.

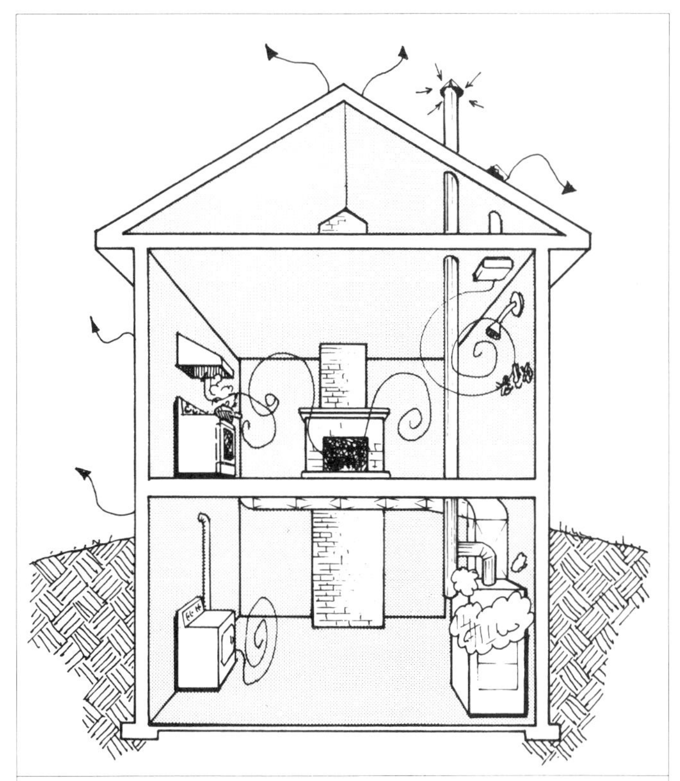

There are many appliances that can compete for household air.

Some of the signs of combustion air problems include:
- back-puffing of the furnace, indicated by soot or staining around the air intake;
- unusual odors from the combustion appliance;
- difficulty starting or maintaining a fire; and
- frequent headaches or nausea.

The ideal solution is to remove the source of the problem. If you are replacing your heating system anyway, consider buying a system that uses little or no combustion and dilution air from inside the house (e.g. a mid- or high-efficiency heating system, an integrated heat pump system, etc.)

Fireplaces are one of the worst culprits for robbing household air. This can be minimized by installing an outside air duct to the firebox and tight-fitting glass doors. Otherwise, a window should be opened a crack when you operate the fireplace.

Oil and gas furnaces may require a free and unobstructed supply of outside air. The size and type of ducting system will vary according to the type of fuel and the location and heating capacity of the appliance. Obtain specific information on all aspects of combustion air supply systems from your local regulatory authority, fuel supplier or heating contractor before work begins.

Part III Other Features

The whole, they say, is greater than the sum of its parts. In no place is this more true than in your household. The tips below are worth their weight in fuel.

Rugs and Curtains

A rug on any floor over a cold part of the house (a garage, a cold cellar, or even a coolish basement) will help to insulate. This will not reduce heat loss very well, but it will make the floor feel warmer.

Despite all of the other work you've done to keep the heat in, your windows still represent a weak link. However, with proper design and use of window insulation you can benefit from solar gain during the day and lessen heat losses at night. In the winter, shut your curtains to reduce heat loss once the sun has gone down, and open them during the day. During the summer, keep curtains and windows closed during the day to help prevent overheating.

Lighting

Using lights efficiently and using efficient lights will help you achieve even greater savings. Lights left on unnecessarily, for example, will only add to your energy costs. Lighting only the specific work area (task lighting) allows a reduction in general lighting levels and will save energy.

There are also a variety of new light technologies available that are ideal for homes. PL lamps are pencil thin fluorescent tubes, bent into a "U" shape to reduce their size and attached to a regular screw-in base; they are ideal for task lighting purposes. A 15W PL lamp may replace a 60W incandescent bulb, and give you 10 times the life span.

Incandescent lights with halogen (PAR lamps) are perfect for track lighting, heat lamps, or exterior floods, and can cut energy use by 30 to 50 per cent compared to regular incandescents.

Appliances

All appliances can be used more efficiently. For example, be careful not to leave the refrigerator and freezer door open longer than necessary or leave electric kettles boiling.

When buying a new major appliance, take the time to compare the rating of its energy use. The **Ener-Guide** rating provides the monthly energy consumption in kilowatt-hours for that particular model. These ratings are clearly listed on the **EnerGuide** labels of refrigerators, freezers, ranges, dish washers, clothes washers, and clothes dryers and can be used to compare one model to another. Alternatively, you can refer to the *EnerGuide Directory* available from Natural Resources Canada which lists the rating of all the models currently on the market. Even though energy-efficient appliances may cost more, paybacks are generally very attractive. Read a *Consumer's Guide to Buying and Using Energy-Efficient Appliances* for more information (see page 107).

General Maintenance

Once you have your house operating efficiently, it is important that you keep it that way. This means setting yourself a schedule and checking to make sure the energy saving measures you've initiated are measuring up. Keep an eye on your heating system, and perform the regular maintenance.

Check for signs of moisture damage or structural deterioration, and take immediate action if you locate any trouble spots. Its a good idea to inspect your attic during the cold months. Extensive frost build-up is a pretty good indication that you've got a moisture problem.

Don't forget, if you look after the little things, the heating bill will look after itself. Make sure your weatherstripping is in place and functioning as it should. Check any caulking you have applied for cracks, and recaulk as necessary. It is important to locate and remedy small problems and not give them a chance to become large problems.

These simple measures should all save heat, and leave you with a healthier, happier home!

9 Dollars and Sense

Introduction

At some point, most homeowners hire a contractor for repairs, retrofit or renovation work. Even ardent "do-it-yourselfers" hire a contractor for work that requires special tools, equipment or expertise. Whether you plan to hire someone for a specialized part of the job, or for the whole retrofit, you'll want to ensure that you're satisfied with the results. Many of the tips in this chapter will save you time and money even if you do all the work yourself. Following a few simple guidelines will ensure that the work goes smoothly, that it meets your expectations, and that you get what you're paying for.

- Be an informed consumer.
- Inspect contractors when they inspect the work.
- Use "smart" scheduling.
- Shop around and compare.
- Get a permit if you need one.
- Provide clear instructions.
- Get a written quote.
- Use a contract.
- Monitor the job.

Keep Informed

The housing renovation industry is one of the fastest growing industries in Canada, accounting for more than half of all residential construction work. It is a complex industry that includes general contractors and many different specialty trades. Like any other industry, housing renovation is constantly changing as businesses and regulatory agencies respond to changing market conditions and consumer demands. The industry is also highly subject to local conditions such as availability of materials, and general level of construction activity. Understanding these factors will help you to choose the right contractor for the job.

There are many different types of contractors who specialize in various aspects of retrofit work.

- Basement damp-proofing and foundation specialists work with concrete and masonry to repair, damp-proof and insulate foundations.
- Blown-in insulation installers use specialized equipment to blow cellulose or mineral fibre insulation into wall cavities or attics.
- Spray-on insulation installers use specialized equipment to spray foam or cellulose insulation.
- Carpenters are the generalists of the construction field. They will frame and insulate walls from the interior or exterior and install the air and vapor barrier. Often, they will also install attic insulation, wood siding and windows and doors.
- Siding specialists are often dealer-installers for one type of vinyl, steel or aluminum siding. Sometimes their business combines siding with soffit and fascia installation or with window and door installation.
- Roofers will repair roofs and replace elements such as flashing and shingles. They may also insulate the roof from the exterior.
- Window and door installers are often dealer-installers for one or more products. In the case of storm windows they sometimes assemble custom-sized units from components supplied by the manufacturer.
- Air sealing specialists concentrate on tightening the air barrier throughout the house, using caulking and weatherstripping. They will often test the house air tightness before and after the work, using a fan-door apparatus.
- Some heating contractors are trained in tune-up methods that go beyond regular furnace cleaning and maintenance.
- Other trades such as electricians, plumbers, drywallers, finish carpenters, tile setters, painters and decorators may also be involved in a retrofit project.
- General contractors are the band leaders of the renovation field. They orchestrate and direct all the other trades but they usually employ their own framers and finish carpenters. If you are hiring one or more trades then you are acting as your own general contractor on the job.

"Smart" Scheduling

Seasonal and local conditions will affect the availability of contractors and materials. Being aware of the conditions in your area will help you to schedule your work for the best time. Many trades work in both new construction and retrofit. If there is a new construction "boom" on in your area, then fewer contractors will be available for smaller retrofit jobs. The construction field is seasonally dependent. When the season has been busy there is a scramble to "close-in" buildings before the snow flies. The fall may not be the best time to look for a contractor to do an exterior retrofit.

Strikes that affect new construction also affect the renovation field. Material shortages can also upset schedules and the normal flow of work.

Smart scheduling, particularly for smaller jobs, can make the job run smoothly and save you a lot of money. Make use of seasonal and local conditions to order materials and hire contractors in the "off season" when materials are on sale and contractors aren't charging a premium. Generally, winter is the slow season in construction but ideal for interior work. An early spring start for outside work will give you a better choice of contractors and lower prices. Some materials such as insulation are subject to seasonal sales as manufacturers and building supply houses clear out last year's inventory. Check with your local building supply house that services contractors. When there is a shortage of materials or a construction boom, it may pay you to postpone your work a few months.

Comparison Shop

Retrofitting can be a very good investment, but it is still important to get the most for your money. Both the price of materials and the rates of contractors can vary considerably. It almost always pays to comparison shop before you go ahead.

The renovation and retrofit industry is known for a high turnover of businesses. Knowing what to look for can help you select contractors who are serious about their business. Conditions will vary among provinces and communities but it's good practice to look for the following certifications and credentials:

- **Business license**: All provinces license companies who sign contracts in the home. Check with your provincial department of consumer affairs. (Some municipalities also require a business license.)
- **Trade membership:** Many types of contractors have their own trade associations; many heating contractors belong to the Heating, Refrigerating and Air Conditioning Institute (HRAI), while the National Energy Conservation Association (NECA) is the association whose members do energy retrofit work on the building envelope. The Canadian Home Builders' Association (CHBA) has renovation councils in many provinces who represent general renovation contractors. Association membership is an indication to the consumer that the contractor is concerned about the quality of work performed by their industry.
- **Better Business Bureau:** These organizations maintain records for any complaints against businesses.
- **Technical qualifications**: In some provinces, trades such as electricians, plumbers and heating contractors are required to have technical certification as well as a business license. In addition, technical update courses are offered by various associations for their members. Check with the association in your area to find out what technical qualifications apply. Ask your prospective contractor what qualifications and technical education they and their workers possess and how regularly they update them.
- **Insurance**: Contractors should carry insurance for their business and the work they do on your home. This should include public liability, property damage, and any damage within the first year after completion of the work.
- **Warranty:** Contractors should always provide a warranty or guarantee on the work performed. In some cases the warranty is provided solely by the contractor while in other trades it is backed by the trade association or third party insured. One such warranty program is now available for energy conservation specialists through the Energy Conservation Contractors Warranty Corporation.
- **Track record**: The best recommendation for any business is a satisfied customer. Ask for recommendations from friends and neighbors and check out recent work performed by the company.

To find the right contractor, start with a list of those recommended by your friends, other tradespeople you've dealt with, or building suppliers. Then make sure they are experienced with the type of work you're proposing and that the size of your job is suited to the scale of their business. A large general contractor who did a successful major renovation for your friends may not be the right choice for a small attic retrofit. Finally, check to see how busy they are; you don't want to be last in line for a contractor who's overbooked.

As you narrow down the list, you should end up with at least three contractors whom you will ask to quote on the job.

Permits, Codes and Standards

While you're at the planning stage, check with the Building Department of your municipality to find out what permits you'll need. This varies from province to province but usually special permits are required for any changes to plumbing, heating and electrical wiring. Building permits are also required for any excavation, additions, changes, or alterations to the walls of a building.

The purpose of the permit process and the related inspections is to ensure that the work on your home meets provincial or municipal requirements for health and safety, and that it is structurally sound. The National Building Code, developed by the Associate Committee on the National Building Code, serves as a model code for the provinces who have the authority to regulate buildings. Membership on this committee includes architects, engineers, academics, quantity surveyors, municipal inspectors, builders and consumers. Many provinces have adopted the National Code in whole or in part and some have a separate code or section of the code governing renovation activities. In some cases the province delegates some of the supervisory role to the municipality.

The building inspector can be a valuable resource for your project. With their many years of experience in the construction field, inspectors can often provide valuable advice on quality of workmanship, local construction practices, suitable materials and other concerns, in addition to ensuring that your work meets the building code.

Often codes will state that products or installation methods must conform to a certain standard. There are three organizations that write standards, and test and certify products affecting residential renovation work.

- The Canadian Standards Association (CSA) is a non-government agency that prepares standards and certifies products such as electrical fixtures.
- The Underwriters Laboratory of Canada (ULC) certifies mainly products related to fire safety.
- The Canadian General Standards Board (CGSB) is a government agency operating as a national standards writing body.

Provide Clear Instructions

Whether you plan to hire just one sub-trade or a general contractor for a major renovation, it's important to be specific about the work to be done. Without clear instructions to all concerned, misunderstandings can occur and you may not get what you thought you bargained for.

Giving clear instructions about what you want and expect is important at three different stages of the job.

- **Planning**: Drawings and specifications may be required for your permit application. Preparing written instructions will also help you to ensure that you've included all essential aspects of the job in your plans.
- **Getting quotes:** Clear directions are necessary to obtain comparable quotes with all the contractors bidding on the same materials and quality of work.
- **Under contract**: Including clear instructions in the contract will help to ensure that there are no misunderstandings while the work is in progress.

The amount of detail in your written instructions will vary depending on the size of the job. Remember, many products are hidden from sight so it is important to clearly spell out what you will get. For a simple attic retrofit you may just need a set of written notes outlining the work. This might include the area to be insulated, the amount and type of insulation and RSI value to be achieved, and directions for installing the air barrier, vapor barrier and any required ventilation. A major renovation or retrofit may require some or all of the following:

- a site plan;
- floor plans;
- elevations of the house exterior;
- a cross section of the house;

– detail drawings; and
– written specifications.

Your local Building Department can advise you on what information is required for the permit. In addition, for major jobs you should include instructions about disposal of building materials, responsibility for permit applications, temporary services if needed, and other details.

Get a Written Quote

For proper comparison, you will need a number of quotes. Using your written instructions will ensure that all the contractors are bidding on the same job. Always insist on a written quote that includes all the details in your set of instructions. It's also wise practice to ask the contractors to supply two or more references of previous clients with their quote. When comparing quotes, double check to make sure that the contractor understood the instructions; a very low bid may mean that the contractor misunderstood the work or substituted lower quality materials and workmanship.

Take the time to evaluate all the quotes and to check the references. Don't let anyone pressure you into signing a contract before you're ready.

Use a Contract

Once you've chosen a contractor, insist on a written contract. The contractor may have a standard printed form but it should be used only as a starting point for negotiations.

While there is no such thing as a standard contract, provincial legislation may specify what must be included in a contract for it to be effective. Check with the Consumer Affairs department. The guidelines below provide an example of what is likely to be required.

- Your name and address and the contractor's. This should include the contractor's full company name, phone number and name of the company's signing officer.
- A detailed description of the work to be done under the contract. (This is where your written instructions and any plans and specifications are included.)
- An itemized price for the work and the terms of payment.
- A statement of any warranty or guarantee on the work to be performed.
- Specific dates for starting the work and completing it.
- Signatures of both parties, with each retaining an original signed copy of the contract.

For smaller jobs the contract need not be a complicated document, but it should include all the items listed above. The amount of detail in the contract should increase with the size of the job. For major renovations and retrofit jobs, a full set of contract documents should be prepared, including plans and detailed specifications. These documents should be checked by your lawyer.

Take the time to review the contract carefully (including all the fine print) and make sure that you understand everything before you sign it. A serious contractor will appreciate your thoroughness and won't mind waiting a day or two.

Read the fine print carefully. Some firms have worded their contracts in such a way that they are not responsible for any problems following the work. Such clauses usually imply that any subsequent problems (e.g. peeling paint, cracks) are related to "hidden construction defects" that existed prior to the retrofit work. This may not necessarily be the case. Stroke out these clauses before you sign.

All verbal agreements should be written into the contract. If there are minor changes you wish to make, write them in on both copies of the contract. They will be valid if initialled by both you and the contractor.

Make Sure You Have Recourse

Insist on a written warranty as part of the contract. Responsible contractors will guarantee their work and assume responsibility for problems that could develop later as a result of their work.

Be wary when asked to pay in advance, or in cash to a salesperson rather than by cheque or money order to the company itself. If a down payment is required, keep it as small as possible. Usually the first payment is only made on delivery of materials to the site. Advance down payments are restricted to situations where the contractor has to place an advance order for large quantities of special materials (such as custom made windows). Payment schedules should be tied to work actually completed and not to arbitrary dates. Any provisions for holdbacks should be written into the contract.

You should also be aware that if your contract is sold to a third party, such as a finance company, you are obligated to them regardless of any complaints you may have about the quality of the work.

Take Extra Precautions with Door-to-Door Salespeople

Be especially careful when dealing with firms who sell their services through door-to-door salespeople. Fly-by-night and unreliable businesses are more likely to depend on door-to-door sales. Insist on seeing the sales representative's card. Check if the company has a permanent office location that you can call if problems occur with the product or installation.

Never sign on impulse. Ask the salesperson to give you a copy of the contract so you may think it over; if this request is refused, it's probably because you are on the verge of closing a bad deal. Shop around to compare quality and prices. Don't be pressured into buying.

If you sign a contract in your home, remember that all provincial statutes regulating direct (door-to-door) sales have a so-called cooling-off provision. This gives you the right to cancel a contract signed with a door-to-door salesperson within a specified number of days. The number of days varies from province to province; check with your provincial consumer bureau. There may be other instances where you can cancel your contract after the cooling-off period is over. Again, check with your provincial consumer bureau.

Monitor Performance

When you hire a contractor it's your responsibility to make sure that the job is done on time and on budget. For smaller jobs this may simply mean being on hand to make sure that materials are the same quality and type as stated in the contract and that the work method follows your written instructions. For larger jobs a daily meeting with the contractor may be required to keep track of the work and resolve any problems as they arise. The payment schedule should be tied to work actually done, not predetermined dates. Where permits are involved, you may wish to tie in progress payments to successful passing of the various building inspections during the course of the job.

Your final payment should be made only some time after the work is totally complete. In most provinces legislation permits you to retain a set percentage of the total cost of the work for a specific period of time. The purpose of this holdback is to help limit your liability should a contractor fail to pay suppliers, workers or sub-trades. The legislation varies from province to province. Your provincial consumer bureau will be able to provide you with details.

Canada Mortgage and Housing Corporation offers a free publication entitled: "How to Hire a Contractor" (publication #NHA-5429). To obtain a copy, contact their nearest regional office, or write to the address on page 33.

Easy Reference

Need More Information?
Check Out our Free Publications

Natural Resources Canada (NRCan) has many publications to help you understand home heating systems, home energy use, transportation efficiency, and to explain what you can do to reduce your energy costs while increasing your comfort.

Want to Draftproof and Reduce your Energy Use?
Keeping the Heat In is a guide to all aspects of home insulation and draftproofing. Whether you plan to do it yourself or hire a contractor, this 107-page book can help make it easier. Fact sheets are also available on air leakage control and moisture problems.

How About Home Heating, Cooling and Ventilation Systems?
If you are interested in a particular energy source, NRCan has booklets on heating with electricity, gas, oil, heat pumps, and wood. Other publications are available on heat recovery ventilators, wood fireplaces, gas fireplaces, air conditioning your home and comparing heating costs.

...and Consumer's Guides?
The Consumer's Guides can help you choose energy-efficient items such as office equipment, household appliances, lighting products, and windows and doors.

...and EnerGuide Directories?
The EnerGuide Program helps you choose energy-using products that use the least amount of energy. The EnerGuide label, which is affixed to major electrical household appliances and room air conditioners, helps you choose the most energy-efficient models. Annual directories list the EnerGuide ratings of major electrical household appliances and room air conditioners.

...and Energy-efficient New Housing?
R-2000 Homes use up to 50 per cent less energy than conventional dwellings. Features include state-of-the-art heating systems, high levels of insulation, use of solar energy, and whole-house ventilation systems which provide continuous fresh air to all rooms. Once completed, R-2000 Homes are subject to third-party testing.

...and Transportation Efficiency?
The *Auto$mart Guide* shows you how to buy, drive and maintain your car to save money and energy.
The *Car Economy Calculator* helps you determine your vehicle's fuel consumption. The annual *Fuel Consumption Guide* lists the fuel consumption ratings of most new vehicles sold in Canada. Information is also available on fuel alternatives to gasoline and diesel (e.g., propane, natural gas, ethanol and methanol).

To receive any of the free publications listed above, please call us or write to

Energy Publications
c/o Canada Communication Group
Ottawa, ON
K1A OS9

Facsimile: (819) 994-1498
Toll-free: 1-800-387-2000
In Ottawa, call 995-2943.

Please allow three weeks for delivery.

For more information on energy efficiency at home and on the road or to order these publications online, visit our Web site at http://eeb-dee.nrcan.gc.ca.

Notes